# Lecture Notes in Mathematics

## 1610

Editors:
A. Dold, Heidelberg
F. Takens, Groningen

# Springer

*Berlin*
*Heidelberg*
*New York*
*Barcelona*
*Budapest*
*Hong Kong*
*London*
*Milan*
*Paris*
*Santa Clara*
*Singapore*
*Tokyo*

Ali Süleyman Üstünel

# An Introduction to
# Analysis on Wiener Space

 Springer

Author

Ali Süleyman Üstünel
ENST, Dépt. Réseaux
46, rue Barrault
F-75013 Paris, France
e-mail: ustunel@res.enst.fr

Cataloging-in-Publication Data applied for

Die Deutsche Bibliothek - CIP-Einheitsaufnahme

**Üstünel, Ali Süleyman:**
An introduction to analysis on Wiener space / Ali Süleyman
Üstünel. - Berlin ; Heidelberg ; New York ; Barcelona ;
Budapest ; Hong Kong ; London ; Milan ; Paris ; Tokyo :
Springer, 1995
  (Lecture notes in mathematics ; Vol 1610)
  ISBN 3-540-60170-8
NE: GT

Mathematics Subject Classification (1991): 60H07, 60H05, 60H15, 46F25, 81T,
81Q

ISBN 3-540-60170-8 Springer-Verlag Berlin Heidelberg New York

© Springer-Verlag Berlin Heidelberg 1995
Printed in Germany

Typesetting: Camera-ready T$_E$X output by the author
SPIN: 10479544      46/3142-543210 - Printed on acid-free paper

Ankara Fen Lisesi 67-70 mezunlarına .

# Introduction

The following pages are the notes from a seminar that I gave during the spring and some portion of the summer of 1993 at the Mathematics Institute of Oslo University. The aim of the seminars was to give a rapid but rigorous introduction for the graduate students to Analysis on Wiener space, a subject which has grown up very quickly these recent years under the impulse of the Stochastic Calculus of Variations of Paul Malliavin (cf. [12]).

Although some concepts are in the preliminaries, I assumed that the students had already acquired the notions of stochastic calculus with semimartingales, Brownian motion and some rudiments of the theory of Markov processes. A small portion of the material exposed is our own research, in particular, with Moshe Zakai. The rest has been taken from the works listed in the bibliography.

The first chapter deals with the definition of the (so-called) Gross-Sobolev derivative and the Ornstein-Uhlenbeck operator which are indispensable tools of the analysis on Wiener space. In the second chapter we begin the proof of the Meyer inequalities, for which the hypercontractivity property of the Ornstein-Uhlenbeck semigroup is needed. We expose this last topic in the third chapter, then come back to Meyer inequalities, and complete their proof in chapter IV. Different applications are given in next two chapters. In the seventh chapter we study the independence of some Wiener functionals with the previously developed tools. The chapter VIII is devoted to some series of moment inequalities which are important for applications like large deviations, stochastic differential equations, etc. In the last chapter we expose the contractive version of Ramer's theorem as another example of the applications of moment inequalities developed in the preceding chapter.

During my visit to Oslo, I had the chance of having an ideal environment for working and a very attentive audience in the seminars. These notes have particularly profited from the serious criticism of my colleagues and friends Bernt Øksendal, Tom Lindstrøm, Ya-Zhong Hu, and the graduate students of the Mathematics department. It remains for me to express my gratitude also to Nina Haraldsson for her careful typing, and, last but not least, to Laurent Decreusefond for correcting so many errors .

Ali Süleyman Üstünel

# Contents

x

.

# Preliminaries

This chapter is devoted to the basic results about the Wiener measure, Brownian motion, construction of the Ito stochastic integral and the chaos decomposition associated to it.

## 1 The Brownian Motion and the Wiener Measure

**1)**  Let $W = C_0([0,1])$, $\omega \in W$, $t \in [0,1]$, define $W_t(\omega) = \omega(t)$ (the coordinate functional). If we note by $\mathcal{B}_t = \sigma\{W_s; s \leq t\}$, then there is one and only one measure $\mu$ on $W$ such that

i) $\mu\{W_0(\omega) = 0\} = 1$,

ii) $\forall f \in C_b^\infty(\mathbf{R})$, the stochastic process process

$$(t,\omega) \mapsto f(W_t(\omega)) - \frac{1}{2}\int_0^t f''(W_s(\omega))ds$$

is a $(\mathcal{B}_t, \mu)$-martingale. $\mu$ is called the Wiener measure.

**2)**  From the construction we see that for $t > s$,

$$E_\mu[\exp i\alpha(W_t - W_s)|\mathcal{B}_s] = \exp -\alpha^2(t - s),$$

hence $(t,\omega) \mapsto W_t(\omega)$ is a continuous additive process (i.e.,a process with independent increments) and $(W_t; t \in [0,1])$ is also a continous martingale.

**3) Stochastic Integration**

Let $K : W \times [0,1] \to \mathbf{R}$ be a step process :

$$K_t(\omega) = \sum_{i=1}^n a_i(\omega) \cdot 1_{[t_i, t_{i+1}[}(t), \qquad a_i(\omega) \in L^2(\mathcal{B}_{t_i}).$$

Define

$$I(K) = \int_0^1 K_s dW_s(\omega)$$

as

$$\sum_{i=1}^{n} a_i(\omega) \cdot (W_{t_{i+1}}(\omega) - W_{t_i}(\omega)).$$

Then we have

$$E\left[\left(\int_0^1 K_s dW_s\right)^2\right] = E \int_0^1 K_s^2 ds,$$

i.e. $I$ is an isometry from the adapted step processes into $L^2(\mu)$, hence it has a unique extension as an isometry from

$$L^2([0,1] \times W, \mathcal{A}, dt \times d\mu) \xrightarrow{I} L^2(\mu)$$

where $\mathcal{A}$ denotes the sigma algebra on $[0,1] \times W$ generated by the adapted, left (or right) continuous processes. $I(K)$ is called the stochastic integral of $K$ and it is denoted as $\int_0^1 K_s dW_s$. If we define

$$I_t(K) = \int_0^t K_s dW_s$$

as

$$\int_0^1 1_{[0,t]}(s)K_s dW_s,$$

it is easy to see that the stochastic process $t \mapsto I_t(K)$ is a continuous, square integrable martingale. With some localization techniques using stopping times, $I$ can be extended to any adapted process $K$ such that $\int_0^1 K_s^2(\omega)ds < \infty$ a.s. In this case the process $t \mapsto I_t(K)$ becomes a local martingale, i.e., there exists a sequence of stopping times increasing to one, say $(T_n, n \in \mathbf{N})$ such that the process $t \mapsto I_{t \wedge T_n}(K)$ is a (square integrable) martingale.

**Application: Ito formula**   We have following important applications of the stochastic integration:

a) If $f \in C^2(\mathbf{R})$ and $M_t = \int_0^t K_r dW_r$, then

$$f(M_t) = f(0) + \int_0^t f'(M_s)K_s dW_s + \frac{1}{2}\int_0^t f''(M_s)K_s^2 ds.$$

b)

$$\mathcal{E}_t(I(h)) = \exp\left(\int_0^t h_s dW_s - \frac{1}{2}\int_0^t h_s^2 ds\right)$$

is a martingale for any $h \in L^2[0,1]$.

## 4) Alternative constructions of the Brownian motion and the Wiener measure

**A)**  Let $(\gamma_i; i \in \mathbf{N})$ be an independent sequence of $N_1(0,1)$ Gaussian random variables. Let $(g_i)$ be a complete, orthonormal basis of $L^2[0,1]$. Then $W_t$ defined by

$$W_t(\omega) = \sum_{i=1}^{\infty} \gamma_i(\omega) \cdot \int_0^t g_i(s)ds$$

is a Brownian motion.

**Remark:**  If $(g_i; i \in \mathbf{N})$ is a complete, orthonormal basis of $L^2([0,1])$, then $\left( \int_0^{\cdot} g_i(s)ds; i \in \mathbf{N} \right)$ is a complete orthonormal basis of $H([0,1])$ (i. e., the first order Sobolev functionals on $[0,1]$).

**B)**  Let $(\Omega, \mathcal{F}, \mathbf{P})$ be any abstract probability space and let $H$ be any separable Hilbert space. If $L : H \to L^2(\Omega, \mathcal{F}, \mathbf{P})$ is a linear operator such that for any $h \in H$, $E[\exp iL(h)] = \exp -\frac{1}{2}|h|_H^2$, then there exists a Banach space with dense injection $H \xrightarrow{j} W$ dense, hence $W^* \xrightarrow{j^*} H$ is also dense and a probability measure $\mu$ on $W$ such that

$$\int \exp\langle \omega^*, \omega \rangle d\mu(\omega) = \exp -\frac{1}{2} \mid j^*(\omega^*) \mid_H^2$$

and

$$L(j^*(\omega^*))(\omega) = \langle \omega^*, \omega \rangle$$

almost surely. $(W, H, \mu)$ is called an Abstract Wiener space and $\mu$ is the Wiener measure. If $H([0,1]) = \{h : h(t) = \int_0^t \dot{h}(s)ds, |h|_H = |\dot{h}|_{L^2[0,1]}\}$ then $\mu$ is the classical Wiener measure and $W$ can be taken as $C_0([0,1])$.

**Remark:**  In the case of the classical Wiener space, any element $\lambda$ of $W^*$ is a signed measure on $[0,1]$, and its image in $H = H([0,1])$ can be represented as $j^*(\lambda)(t) = \int_0^t \lambda([s,1])ds$. In fact, we have for any $h \in H$

$$
\begin{aligned}
(j^*(\lambda), h) &= <\lambda, j(h)> \\
&= \int_0^1 h(s)\lambda(ds) \\
&= h(1)\lambda([0,1]) - \int_0^1 \lambda([0,s])\dot{h}(s)ds \\
&= \int_0^1 (\lambda([0,1]) - \lambda([0,s]))\dot{h}(s)ds \\
&= \int_0^1 \lambda([s,1])\dot{h}(s)ds.
\end{aligned}
$$

**5) Let us come back to the classical Wiener space:**

i) It follows from the martingale convergence theorem and the monotone class theorem that the set of random variables

$$\{f(W_{t_1}, \ldots, W_{t_n}); t_i \in [0,1], f \in \mathcal{S}(\mathbf{R}^n); n \in \mathbf{N}\}$$

is dense in $L^2(\mu)$, where $\mathcal{S}(\mathbf{R}^n)$ denotes the space of infinitely differentiable, rapidly decreasing functions on $\mathbf{R}^n$.

ii) It follows from (i), via the Fourier transform that the linear span of the set $\{\exp \int_0^1 h_s dW_s - \frac{1}{2} \int_0^1 h_s^2 ds; h \in L^2([0,1])\}$ is dense in $L^2(\mu)$.

iii) Because of the analyticity of the characteristic function of the Wiener measure, the elements of the set in (ii) can be approached by the polynomials, hence the polynomials are dense in $L^2(\mu)$.

**5.1 Cameron-Martin Theorem:**

For any bounded Borel measurable function $F$, $h \in L^2[0,1]$, we have

$$E_\mu[F(w + \int_0^\bullet h_s ds) \cdot \exp[-\int_0^1 h_s dW_s - \frac{1}{2} \int_0^1 h_s^2 ds]] = E_\mu[F].$$

This means that the process $W_t(\omega) + \int_0^t h_s ds$ is again a Brownian motion under the new probability measure

$$\exp(-\int_0^1 h_s dW_s - \frac{1}{2} \int_0^1 h_s^2 ds) d\mu.$$

**Proof:** It is sufficient to show that the new probability has the same characteristic function as $\mu$: if $x^* \in W^*$, then $x^*$ is a measure on $[0,1]$ and

$$
\begin{aligned}
_{W^*}\langle x^*, w \rangle_W &= \int_0^1 W_s(\omega) x^*(ds) \\
&= W_t(\omega) \cdot x^*([0,t])\Big|_0^1 - \int_0^1 x^*([0,t]) dW_t(\omega) \\
&= W_1 x^*([0,1]) - \int_0^1 x^*([0,t]).dW_t \\
&= \int_0^1 x^*(]t,1]) dW_t.
\end{aligned}
$$

Consequently

$$E[\exp i \int_0^1 x^*([t, 1])dW_t(w + \int_0^\bullet h_s ds) \cdot \mathcal{E}(-I(h))]$$

$$= E[\exp i \int_0^1 x^*([t, 1])dW_t + i \int_0^1 x^*([t, 1])h_t dt - \int_0^1 h_t dW_t - \frac{1}{2} \int_0^1 h_t^2 dt]$$

$$= E[\exp i \int_0^1 (ix^*([t, 1]) - h_t)dW_t \cdot \exp i \int_0^1 x^*([t, 1])h_t dt - \frac{1}{2} \int_0^1 h_t^2 dt]$$

$$= \exp \frac{1}{2} \int_0^1 (ix^*([t, 1]) - h_t)^2 dt + i \int_0^1 x^*([t, 1])h_t dt - \frac{1}{2} \int_0^1 h_t^2 dt$$

$$= \exp -\frac{1}{2} \int_0^1 (x^*([t, 1]))^2 dt$$

$$= \exp -\frac{1}{2} | j(x^*) |_H^2 .$$

QED

**Corollary (Paul Lévy's Theorem )** Suppose that $(M_t)$ is a continuous martingale such that $M_0 = 0$, $M_t^2 - t$ is again a martingale. Then $(M_t)$ is a Brownian motion.

**Proof:** We have the Ito formula

$$f(M_t) = f(0) + \int_0^t f'(M_s) \cdot dM_s + \frac{1}{2} \int_0^t f''(M_s) \cdot ds .$$

Hence the law of $\{M_t : t \in [0, 1]\}$ is $\mu$. QED

**5.2 The Ito Representation Theorem:**

Any $\varphi \in L^2(\mu)$ can be represented as

$$\varphi = E[\varphi] + \int_0^1 K_s dW_s$$

where $K \in L^2([0, 1] \times W)$, adapted.

**Proof:** Since the Wick exponentials

$$\mathcal{E}(I(h)) = \exp \int_0^1 h_s dW_s - 1/2 \int_0^1 h_s^2 ds$$

can be represented as claimed, the proof follows by density. QED

### 5.3 Wiener chaos representation

Let $K_1 = \int_0^1 h_s dW_s$, $h \in L^2([0,1])$. Then, from the Ito formula, we can write

$$
K_1^p = p \int_0^1 K_s^{p-1} h_s dW_s + \frac{p(p-1)}{2} \int_0^1 K_s^{p-2} h_s^2 ds
$$

$$
= p \int_0^1 \left[ (p-1) \int_0^{t_1} K_{t_2}^{p-2} h_{t_2} dW_{t_2} + \frac{(p-1)(p-1)}{2} \int_0^{t_1} K_{t_2}^{p-3} h_{t_2}^2 dt_2 \right] dW_{t_1}
$$

$$
+ \cdots
$$

iterating this procedure we end up on one hand with $K_{t_p}^0 = 1$, on the other hand with the multiple integrals of deterministic integrands of the type

$$
J_p = \int\limits_{0 < t_p < t_{p-1} < \cdots \le t_1 < 1} h_{t_1} h_{t_2} \ldots h_{t_p} \, dW_{t_1}^{i_1} \ldots dW_{t_p}^{i_p},
$$

$i_j = 0$ or $1$ with $dW_t^0 = dt$ and $dW_t^1 = dW_t$.

Let now $\varphi \in L^2(\mu)$, then we have from the Ito representation theorem

$$
\varphi = E[\varphi] + \int_0^1 K_s dW_s
$$

by iterating the same procedure for the integrand of the above stochastic integral:

$$
\varphi = E[\varphi] + \int_0^1 E[K_s] dW_s + \int_0^1 \int_0^{t_1} E[K_{t_1,t_2}^{1,2}] dW_{t_2} dW_{t_1} +
$$

$$
+ \int_0^1 \int_0^{t_1} \int_0^{t_2} K_{t_1 t_2 t_3}^{1,2,3} dW_{t_3} dW_{t_2} dW_{t_1} .
$$

After $N$ iterations we end up with

$$
\varphi = \sum_0^N J_p(K^p) + \varphi_{N+1}
$$

and each element of the sum is orthogonal to the other one. Hence $(\varphi_N; N \in \mathbf{N})$ is bounded in $L^2(\mu)$. Let $(\varphi_{N_k})$ be a weakly convergent subsequence and $\varphi_\infty = \lim_{k \to \infty} \varphi_{N_k}$. Then it is easy from the first part that $\varphi_\infty$ is orthogonal to the polynomials, therefore $\varphi_\infty = 0$ and $w - \lim_{N \to \infty} \sum_0^N J_p(K_p)$ exists, moreover $\sup_N \sum_1^N \|J_p(K_p)\|_2^2 < \infty$, hence $\sum_1^\infty J_p(K_p)$ converges in $L^2(\mu)$. Let now $\widehat{K}_p$ be an element of $\widehat{L}^2[0,1]^p$ (i.e. symmetric), defined as $\widehat{K}_p = K_p$ on $C_p = \{t_1 < \cdots < t_p\}$. We define $I_p(\widehat{K}_p) = p! J_p(K_p)$ in such a way that

$$
E[|I_p(\widehat{K}_p)|^2] = (p!)^2 \int_{C_p} K_p^2 dt_1 \ldots dt_p = p! \int_{[0,1]^p} |\widehat{K}_p|^2 dt_1 \ldots dt_p .
$$

Let $\varphi_p = \dfrac{\widehat{K}_p}{p!}$, then we have

$$\varphi = E[\varphi] + \sum_{1}^{\infty} I_p(\varphi_p) \qquad \text{(Wiener chaos decomposition)}$$

# Chapter I

# Gross-Sobolev Derivative, Divergence and Ornstein-Uhlenbeck Operator

## Motivations

Let $W = C_0([0,1], \mathbf{R}^d)$ be the classical Wiener space equipped with $\mu$ the Wiener measure. We want to construct on $W$ a Sobolev type analysis in such a way that we can apply it to the random variables that we encounter in the applications. Mainly we want to construct a differentiation operator and to be able to apply it to practical examples. The Fréchet derivative is not satisfactory. In fact the most frequently encountered Wiener functionals, as the multiple (or single) Wiener integrals or the solutions of stochastic differential equations with smooth coefficients are not even continuous with respect to the Fréchet norm of the Wiener space. Therefore, what we need is in fact to define a derivative on the $L^p(\mu)$-spaces of random variables, but in general, to be able to do this, we need the following property which is essential: if $F, G \in L^p(\mu)$, and if we want to define their directional derivative, in the direction, say $\tilde{w} \in W$, we write $\frac{d}{dt}F(w + t\tilde{w})|_{t=0}$ and $\frac{d}{dt}G(w + t\tilde{w})|_{t=0}$. If $F = G$ $\mu$-a.s., it is natural to ask that their derivatives are also equal a.s. For this, the only way is to choose $\tilde{w}$ in some specific subspace of $W$, namely, the Cameron-Martin space $H$:

$$H = \left\{ h : [0,1] \to \mathbf{R}^d / h(t) = \int_0^t \dot{h}(s)ds, \quad |h|_H^2 = \int_0^1 |\dot{h}(s)|^2 ds \right\}.$$

In fact, the theorem of Cameron-Martin says that for any $F \in L^p(\mu)$, $p > 1$,

$h \in H$

$$E_\mu[F(w+h)\exp[-\int_0^1 \dot{h}(s) \cdot dW_s - \tfrac{1}{2}|h|^2_{H_1}]] = E_\mu[F],$$

or equivalently

$$E_\mu[F(w+h)] = E[F(w) \cdot \exp\int_0^1 \dot{h}_s \cdot dW_s - \tfrac{1}{2}|h|^2_{H_1}].$$

That is to say, if $F = G$ a.s., then $F(\cdot + h) = G(\cdot + h)$ a.s. for all $h \in H$.

# 1   The Construction of $\nabla$ and its properties

If $F : W \to \mathbf{R}$ is a function of the following type (called cylindrical ):

$$F(w) = f(W_{t_1}(w), \dots, W_{t_n}(w)), \qquad f \in \mathcal{S}(\mathbf{R}^n),$$

we define, for $h \in H$,

$$\nabla_h F(w) = \frac{d}{d\lambda} F(w + \lambda h)|_{\lambda=0}.$$

Noting that $W_t(w + h) = W_t(w) + h(t)$, we obtain

$$\nabla_h F(w) = \sum_{i=1}^n \partial_i f(W_{t_1}(w), \dots, W_{t_n}(w))h(t_i),$$

in particular

$$\nabla_h W_t(w) = h(t) = \int_0^t \dot{h}(s)ds = \int_0^1 1_{[0,t]}(s) \; \dot{h}(s)ds.$$

If we denote by $U_t$ the element of $H$ defined as $U_t(s) = \int_0^s 1_{[0,t]}(r)dr$, we have $\nabla_h W_t(\omega) = (U_t, h)_H$. Looking at the linear map $h \mapsto \nabla_h F(\omega)$ we see that it defines a random element with values in $H$, i.e. $\nabla F$ is an $H$-valued random variable. Now we can prove:

**Prop. I.1:**   $\nabla$ is a closable operator on any $L^p(\mu)$ $(p > 1)$.

**Proof:**   This means that if $(F_n : n \in \mathbf{N})$ are cylindrical functions on $W$, such that $F_n \to 0$ in $L^p(\mu)$ and if $(\nabla F_n; n \in \mathbf{N})$ is Cauchy in $L^p(\mu, H)$, then its limit is zero. Hence suppose that $\nabla F_n \to \xi$ in $L^p(\mu; H)$.

To prove $\xi = 0$ $\mu$-a.s., we use the Cameron-Martin theorem: Let $\varphi$ be any cylindrical function. Since such $\varphi$'s are dense in $L^p(\mu)$, it is sufficient to prove that $E[(\xi, h)_H \cdot \varphi] = 0$ a.s. for any $h \in H$. But we have

$$E[(\nabla F_n, h)\varphi] = \frac{d}{d\lambda} E[F_n(w + \lambda h) \cdot \varphi]|_{\lambda=0}$$

$$= \frac{d}{d\lambda} E[F_n(\omega)\varphi(w - \lambda h) \exp(\lambda \int_0^1 \dot{h}(s) dW_s - \frac{\lambda^2}{2} \int_0^1 |\dot{h}_s|^2 ds)]|_{\lambda=0}$$

$$= E[F_n(w)(-\nabla_h \varphi(w) + \varphi(w) \int_0^1 \dot{h}(s) dW_s)] \xrightarrow[n\to\infty]{} 0$$

by the fact that $F_n \to 0$ in $L^p(\mu)$. QED

This result tells us that we can define $L^p$-domain of $\nabla$, denoted by $\mathrm{Dom}_p(\nabla)$ as

**Definition:** $F \in \mathrm{Dom}_p(\nabla)$ if and only if there exists a sequence $(F_n; n \in \mathbf{N})$ of cylindrical functions such that $F_n \to F$ in $L^p$ and $(\nabla F_n)$ is Cauchy in $L^p(\mu, H)$. Then we define

$$\nabla F = \lim_{n\to\infty} \nabla F_n.$$

The extended operator $\nabla$ is called **Gross-Sobolev derivative** .

We will denote by $D_{p,1}$ the linear space $\mathrm{Dom}_p(\nabla)$ equipped with the norm $\|F\|_{p,1} = \|F\|_p + \|\nabla F\|_{L^p(\mu,H)}$.

**Remarks:** 1) If $\mathcal{X}$ is a separable Hilbert space we can define $D_{p,1}(\mathcal{X})$ exactly in the same way as before, the only difference is that we take $\mathcal{S}_{\mathcal{X}}$ instead of $\mathcal{S}$, i.e., the rapidly decreasing functions with values in $\mathcal{X}$. Then the same closability result holds (exercise!).
2) Hence we can define $D_{p,k}$ by iteration:

   i) We say that $F \in D_{p,2}$ if $\nabla F \in D_{p,1}(H)$, then write $\nabla^2 F = \nabla(\nabla F)$.

   ii) $F \in D_{p,k}$ if $\nabla^{k-1} F \in D_{p,1}(H^{\otimes(k-1)})$.

3) Note that, for $F \in D_{p,k}$, $\nabla^k F$ is in fact with values $H^{\widehat{\otimes}k}$ (i.e. symmetric tensor product).

4) From the proof we have that if $F \in D_{p,1}$, $h \in H_1$ and $\varphi$ is cylindrical, we have

$$E[\nabla_h F \cdot \varphi] = -E[F \cdot \nabla_h \varphi] + E[I(h) \cdot F \cdot \varphi],$$

where $I(h)$ is the first order Wiener integral of the (Lebesgue) density of $h$. If $\varphi \in D_{q,1}$ ($q^{-1} + p^{-1} = 1$), by a limiting argument, the same relation holds again. Let us note that this limiting procedure shows in fact that if $\nabla F \in L^p(\mu, H)$ then $F.I(h) \in L^p(\mu)$, i.e., $F$ is more than $p$-integrable. This observation gives rise to the logarithmic Sobolev inequality.

## 1.1   Relations with the stochastic integration

Let $\varphi = f(W_{t_1}, \dots, W_{t_n})$, $t_i \leq t$, $f$ smooth. Then we have

$$\nabla_h \varphi(w) = \sum_{i=1}^{n} \partial_i f(W_{t_1}, \dots, W_{t_n}) h(t_i) \, ,$$

hence $\nabla \varphi$ is again a random variable which is $\mathcal{B}_t$-measurable. In fact this property is satisfied by a larger class of Wiener functionals:

**Proposition II.1**   Let $\varphi \in D_{p,1}$, $p > 1$ and suppose that $\varphi$ is $\mathcal{B}_t$-measurable for a given $t \geq 0$. Then $\nabla \varphi$ is also $\mathcal{B}_t$-measurable and furthermore, for any $h \in H_1$, whose support is in $[t, 1]$, $\nabla_h \varphi = (\nabla \varphi, h)_H = 0$ a.s.

**Proof:**   Let $(\varphi_n)$ be a sequence of cylindrical random variable converging to $\varphi$ in $D_{p,1}$. If $\varphi_n$ is of the form $f(W_{t_1}, \dots, W_{t_k})$, it is easy to see that, even if $\varphi_n$ is not $\mathcal{B}_t$-measurable, $E[\varphi_n | \mathcal{B}_t]$ is another cylindrical random variable, say $\theta_n(W_{t_1 \wedge t}, \dots, W_{t_k \wedge t})$. In fact, suppose that $t_k > t$ and $t_1, \dots, t_{k-1} \leq t$. We have

$$E[f(W_{t_1}, \dots, W_{t_k}) | \mathcal{B}_t] = E[f(W_{t_1}, \dots, W_{t_{k-1}}, W_{t_k} - W_t + W_t) | \mathcal{B}_t]$$

$$= \int_{\mathbf{R}} f(W_{t_1}, \dots, W_{t_{k-1}}, W_t + x) p_{t_k - t}(x) dx$$

$$= \theta(W_{t_1}, \dots, W_{t_{k-1}}, W_t) \, ,$$

and $\theta \in \mathcal{S}$ if $f \in \mathcal{S}(\mathbf{R}^k)$, where $p_t$ denotes the heat kernel. Hence we can choose a sequence $(\varphi_n)$ converging to $\varphi$ in $D_{p,1}$ such that $\nabla \varphi_n$ is $\mathcal{B}_t$-measurable for each $n \in \mathbf{N}$. Hence $\nabla \varphi$ is also $\mathcal{B}_t$-measurable.

If $h \in H_1$ has its support in $[t, 1]$, then, for each $n$, we have $\nabla_h \varphi_n = 0$ a.s., because $\nabla \varphi_n$ has its support in $[0, t]$ as one can see from the explicit calculation for $\nabla \varphi_n$. Taking an a.s. convergent subsequence, we see that $\nabla_h \varphi = 0$ a.s. also. QED.

Let now $K$ be a step process:

$$K_t(w) = \sum_{i=1}^{n} a_i(w) 1_{]t_i, t_{i+1}]}(t)$$

where $a_i \in D_{p,1}$ and $\mathcal{B}_{t_i}$-measurable for any $i$. Then we have

$$\int_0^1 K_s dW_s = \sum_i a_i (W_{t_{i+1}} - W_{t_i})$$

and

$$\nabla_h \int_0^1 K_s dW_s = \sum_1^n \nabla_h a_i (W_{t_{i+1}} - W_{t_i}) + a_i (h(t_{i+1}) - h(t_i))$$

$$= \int_0^1 \nabla_h K_s dW_s + \int_0^1 K_s \dot{h}(s) ds \, .$$

Hence

$$\left| \nabla \int_0^1 K_s dW_s \right|_H^2 \leq 2 \left\{ \left| \int_0^1 \nabla K_s dW_s \right|_H^2 + \int_0^1 |K_s|^2 ds \right\}$$

and

$$E \left[ \left( \left| \nabla \int_0^1 K_s dW_s \right|_H^2 \right)^{p/2} \right] \leq 2^p E \left[ \left( \left| \int_0^1 \nabla K_s dW_s \right|_H \right. \right.$$

$$\left. \left. + \int_0^1 |K_s|^2 ds \right)^{p/2} \right] .$$

Using the Burkholder-Davis-Gundy inequality for the Hilbert space valued martingales, the above quantity is majorized by

$$2 c_p E \left( \left[ \left( \int_0^1 |\nabla K_s|_H^2 ds \right)^{p/2} \right] + E \left[ \left( \int_0^1 |K_s|^2 ds \right)^{p/2} \right] \right)$$

$$= \tilde{c}_p \|\nabla \tilde{K}\|_{L^p(\mu, H \otimes H)}^p + \|\tilde{K}\|_{L^p(\mu, H)} , \quad \text{where } \tilde{K}. = \int_0^. K_r dr .$$

Thanks to this majoration, we have proved:

**Proposition II.2** Let $\tilde{K} \in D_{p,1}(H)$ such that $K_t = \frac{d\tilde{K}(t)}{dt}$ be $\mathcal{B}_t$-measurable for almost all $t$. Then we have

$$\boxed{\nabla \int_0^1 K_s dW_s = \int_0^1 \nabla . K_s dW_s + \tilde{K} \quad \text{a.s.}}$$

**Corollary 1:** If $\varphi = I_n(f_n)$, $f_n \in \hat{L}^2([0,1]^n)$, then we have, for $h \in H_1$,

$$\nabla_h I_n(f_n) = n \int_{[0,1]^n} f(t_1, \ldots, t_n) dW_{t_1}, \ldots, dW_{t_{n-1}} . \dot{h}(t_n) . dt_n .$$

**Proof:** Apply the above proposition $n$-times to the case in which, first $f_n$ is $C^\infty([0,1]^n)$, then pass to the limit in $L^2(\mu)$.      QED

The following result will be extended in the sequel to much larger classes of random variables:

**Corollary 2:** Let $\varphi : W \to \mathbf{R}$ be analytic in $H$-direction. Then we have

$$\varphi = E[\varphi] + \sum_{n=1}^\infty I_n \left( \frac{E[\nabla^n \varphi]}{n!} \right) ,$$

i.e., the kernel $\varphi_n \in \hat{L}^2[0,1]^n$ of the Wiener chaos decomposition of $\varphi$ is equal to

$$\frac{E[\nabla^n \varphi]}{n!} .$$

**Proof:**   We have, on one hand, for any $h \in H$,

$$E[\varphi(w+h)] = E\left[\varphi \cdot \exp \int_0^1 h_s dW_s - \tfrac{1}{2} \int_0^1 h_s^2 ds\right] = E[\varphi \cdot \mathcal{E}(I(h))].$$

On the other hand, from Taylor's formula:

$$
\begin{aligned}
E[\varphi(w+h)] &= E[\varphi] + \sum_1^\infty E\left[\frac{(\nabla^n \varphi(w), h^{\otimes n})}{n!}\right] \\
&= E[\varphi] + \sum_1^\infty \frac{1}{n!}(E[\nabla^n \varphi], h^{\otimes n})_{H^{\otimes n}} \\
&= E[\varphi] + \sum_1^\infty \frac{1}{n!} \frac{E[I_n(E[\nabla^n \varphi]) \cdot I_n(h^{\otimes n})]}{n!} \\
&= E[\varphi] + \sum_1^\infty E\left[\frac{I_n(E[\nabla^n \varphi])}{n!} \frac{I_n(h^{\otimes n})}{n!}\right]
\end{aligned}
$$

hence, from the symmetry, we have

$$I_n(\varphi_n) = \tfrac{1}{n!} I_n(E[\nabla^n \varphi]),$$

where we have used the notation $I_1(h) = I(h) = \int_0^1 h_s dW_s$ and

$$I_n(\varphi_n) = \int_{[0,1]^n} \frac{\partial^n \varphi_n}{\partial t_1 \dots \partial t_n}(t_1, \dots, t_n) dW_{t_1} \dots dW_{t_n}.$$

$$\text{QED}$$

**Definition II.1:**   Let $\xi : W \longrightarrow H$ be a random variable. We say that $\xi \in \mathrm{Dom}_p(\delta)$, if for any $\varphi \in D_{q,1}$ $(q^{-1} + p^{-1} = 1)$, we have

$$E[(\nabla \varphi, \xi)_H] \leq c_{p,q}(\xi) \cdot \|\varphi\|_q,$$

and in this case we define $\delta \xi$ by

$$E[\delta \xi \cdot \varphi] = E[(\xi, \nabla \varphi)_H],$$

i.e., $\delta = \nabla^*$ with respect to the measure $\mu$, it is called the divergence operator (for the emergence of this operator cf. [10] and [7] the references there). Let us give some properties of it:

1.)   Let $a : W \longrightarrow \mathbf{R}$ be "smooth", $\xi \in \mathrm{Dom}_p(\delta)$. Then we have, for any $\varphi \in D_{q,1}$,

$$
\begin{aligned}
E[\delta(a\xi)\varphi] &= E[(a\xi, \nabla\varphi)] \\
&= E[(\xi, a\nabla\varphi)] \\
&= E[(\xi, \nabla(a\varphi) - \varphi \cdot \nabla a)] \\
&= E[\delta\xi \cdot a\varphi - \varphi \cdot (\nabla a, \xi)],
\end{aligned}
$$

hence

$$\delta(a\xi) = a\delta\xi - (\nabla a, \xi).$$

2.) Let $h \in H_1$, then we pretend that

$$\delta h = \int_0^1 h(s)dW_s.$$

To see this, it is sufficient to test this relation on the exponential martingales: if $k \in H_1$, we have

$$E[\delta h . \exp \cdot \int_0^1 k_s dW_s - \tfrac{1}{2} \int_0^1 k_s^2 ds] =$$
$$= E[(h, \nabla \mathcal{E}(I(k))_{H_1})]$$
$$= E[(h, k).\mathcal{E}(I(k))]$$
$$= (h, k)_{H_1} .$$

On the other hand, supposing first $h \in W^*$,

$$E[I(h).\mathcal{E}(I(k))] = E[I(h)(w + k)]$$
$$= E[I(h)] + (h, k)_{H_1}$$
$$= (h, k)_{H_1} .$$

Hence in particular, if we denote by $\tilde{1}_{[s,t]}$ the element of $H$ such that $\tilde{1}_{[s,t]}(r) = \int_0^r 1_{[s,t]}(u)du$, we have that

$$\boxed{\delta(\tilde{1}_{[s,t]}) = W_t - W_s .}$$

3.) Let now $K$ be a step process

$$K_t(v) = \sum_1^n a_i(w).1_{[t_i,t_{i+1}[}(t) ,$$

where $a_i \in D_{p,1}$ and $\mathcal{B}_{t_i}$-measurable for each $i$. Let $\tilde{K}$ be $\int_0^\cdot K_s ds$. Then from the property 1, we have

$$\delta \tilde{K} = \delta\left( \sum_1^n a_i . \tilde{1}_{[t_i,t_{i+1}[} \right) = \sum_1^n \left\{ a_i \delta(\tilde{1}_{[t_i,t_{i+1}[}) - (\nabla a_i, \tilde{1}_{[t_i,t_{i+1}[}) \right\} .$$

From the property 2., we have $\delta(\tilde{1}_{[t_i,t_{i+1}[}) = W_{t_{i+1}} - W_{t_i}$, furthermore, from the proposition II.1, the support of $\nabla a_i$ is in $[0, t_i]$, consequently, we obtain

$$\delta \tilde{K} = \sum_{i=1}^n a_i(W_{t_{i+1}} - W_{t_i}) = \int_0^1 K_s dW_s .$$

Hence we have the important result which says, with some abuse of notation that

**Theorem II.1:** $\text{Dom}_p(\delta)$ $(p > 1)$ contains the adapted stochastic processes (in fact their primitives) such that

$$E\left[\left(\int_0^1 K_s^2 ds\right)^{p/2}\right] < \infty$$

and on this class $\delta$ coincides with the Ito stochastic integral.

**Remark:** To be translated as: the stochastic integral of $K$ is being equal to the divergence of $\tilde{K}$!

We will come back to the notion of divergence later.

# 2  The Ornstein-Uhlenbeck Operator

For a nice function $f$ on $W$, $t \geq 0$, we define

$$P_t f(x) = \int_W f(e^{-t}x + \sqrt{1 - e^{-2t}}\, y)\mu(dy),$$

this expression for $P_t$ is called Mehler's formula. Since $\mu(dx)\mu(dy)$ is invariant under the rotations of $W \times W$, i.e., $(\mu \times \mu)(dx, dy)$ is invariant under the transformation

$$T_t(x, y) = (xe^{-t} + y(1 - e^{-2t})^{1/2}, x(1 - e^{-2t})^{1/2} - ye^{-t}),$$

we have obviously

$$
\begin{aligned}
\|P_t f(x)\|_{L^p(\mu)}^p &\leq \iint |(f \otimes 1)(T_t(x, y))|^p \mu(dx)\mu(dy) \\
&= \iint |(f \otimes 1)(x, y)|^p \mu(dx)\mu(dy) \\
&= \int |f(x)|^p \mu(dx),
\end{aligned}
$$

for any $p \geq 1$, $\|P_t f\|_{L^p} \leq \|f\|_{L^p}$; hence also for $p = \infty$ by duality. A straightforward calculation gives that, for any $h \in H \cap W^*$ $(= W^*)$,

$$
\begin{aligned}
P_t(\mathcal{E}(I(h)) &= \mathcal{E}(e^{-t}I(h)) \\
&= \sum_{n=0}^{\infty} e^{-nt}\frac{I_n(h^{\otimes n})}{n!}.
\end{aligned}
$$

Hence, by homogeneity, we have

$$P_t(I_n(h^{\otimes n})) = e^{-nt}I_n(h^{\otimes n})$$

and by density, we obtain

$$P_t I_n(f_n) = e^{-nt}I_n(f_n),$$

for any $f_n \in \hat{L}^2([0,1]^n)$. Consequently $P_s \circ P_t = P_{s+t}$, i.e., $(P_t)$ is a measure preserving Markov semi-group. Its infinitesimal generator is denoted by $-\mathcal{L}$ and is $\mathcal{L}$ is called the Ornstein-Uhlenbeck or the number operator. Evidently, we have $\mathcal{L}I_n(f_n) = nJ_n(f_n)$; i.e., the Wiener chaos are its eigenspace. From the definition, it follows directly that (for $a_i$ being $\mathcal{F}_{t_i}$-measurable)

$$P_t\left(\sum a_i(W_{t_i+1} - W_{t_i})\right) = e^{-t}\sum(P_t a_i)(W_{t_i+1} - W_{t_i}),$$

that is to say

$$P_t \int_0^1 H_s dW_s = e^{-t}\int_0^1 P_t H_s dW_s,$$

and by differentiation

$$\mathcal{L}\int_0^1 H_s dW_s = \int_0^1 (I + \mathcal{L})H_s dW_s.$$

Also we have

$$\nabla P_t \varphi = e^{-t}P_t \nabla \varphi.$$

**Lemma:** We have $\delta \circ \nabla = \mathcal{L}$, where $\delta$ is the divergence operator (sometimes it is also called Hitsuda-Ramer-Skorohod integral).

**Proof:** Let $\varphi = \mathcal{E}(I(h))$, then

$$
\begin{aligned}
(\delta \circ \nabla)\varphi &= \delta(h.\mathcal{E}(I(h))) \\
&= (I(h) - |h|^2)\mathcal{E}(I(h)) \\
&= \mathcal{L}\mathcal{E}(I(h))
\end{aligned}
$$

QED

Let us define for the smooth functions $\varphi$, a semi-norm

$$\|\varphi\|_{p,k} = \|(I + \mathcal{L})^{k/2}\varphi\|_{L^p(\mu)}.$$

At first glance, these semi-norms (in fact norms), seem different from the one define by $\|\varphi\|_{p,k} = \sum_0^k \|\nabla^j \varphi\|_{L^p(\mu, H^{\otimes j})}$. We will show that they are equivalent. Before that we need

**Proposition** We have the following identity:

$$\delta \circ \nabla = \mathcal{L}.$$

**Proof:**  It is sufficient to prove, for the moment that result, on the exponential martingales; if $h \in H_1$,

$$
\begin{aligned}
\mathcal{L}\mathcal{E}(I(h)) &= \left. -\frac{dP_t}{dt}\mathcal{E}(I(h))\right|_{t=0} \\
&= \left. -\frac{d}{dt}\mathcal{E}(e^{-t}I(h))\right|_{t=0} \\
&= \left. (e^{-t}I(h)) - e^{-2t}|h|_{H_1}^2)\mathcal{E}(e^{-t}I(h))\right|_{t=0} \\
&= (I(h) - |h|^2)\mathcal{E}(I(h)) .
\end{aligned}
$$

On the other hand:

$$
\nabla\mathcal{E}(I(h)) = h \cdot \mathcal{E}(I(h))
$$

and

$$
\begin{aligned}
\delta(\nabla\mathcal{E}(I(h))) &= \delta(h \cdot \mathcal{E}(I(h))) \\
&= \delta h \cdot \mathcal{E}(I(h)) - (\nabla\mathcal{E}(I(h)), h) \\
&= \delta h \mathcal{E}(I(h)) - |h|^2 \mathcal{E}(I(h)) .
\end{aligned}
$$

QED

# Chapter II

# Meyer Inequalities

## Meyer Inequalities and Distributions

Meyer inequalities are essential to control the Sobolev norms defined with the Sobolev derivative with the norms defined via the Ornstein-Uhlenbeck operator. They can be summarized as the equivalence of the two norms defined on the (real-valued) Wiener functionals as

$$|||\phi|||_{p,k} = \sum_{i=0}^{k} ||\nabla^i \phi||_{L^p(\mu, H^{\otimes i})},$$

and

$$||\phi||_{p,k} = ||(I + \mathcal{L})^{k/2} \phi||_{L^p(\mu)},$$

for any $p > 1$ and $k \in \mathbf{N}$. The key point is the continuity property of the Riesz transform on $L^p([0, 2\pi], dx)$, i.e., from a totally analytic origin, although the original proof of P. A. Meyer was probabilistic (cf. [13]). Here we develop the proof suggested by [6].

## 1 Some Preparations

Let $f$ be a function on $[0, 2\pi]$, extended to the whole $\mathbf{R}$ by periodicity. We denote by $\tilde{f}(x)$ the function defined by

$$\tilde{f}(x) = \frac{1}{\pi} p.v. \int_0^{\pi} \frac{f(x+t) - f(x-t)}{2 \tan t/2} dt \quad \text{(principal value)}.$$

then the famous theorem of M. Riesz, cf. [33], asserts that, for any $f \in L^p[0, 2\pi]$, $\tilde{f} \in L^p([0, 2\pi])$, for $1 < p < \infty$ with

$$||\tilde{f}||_p \leq A_p ||f||_p,$$

where $A_p$ is a constant depending only on $p$. Most of the classical functional analysis of the 20-th century has been devoted to extend this result to the case where the function $f$ was taking its values in more abstract spaces than the real line. We will show that our problem also can be reduced to this one.

In fact, the main result that we are going to show will be that

$$\|\nabla (I + \mathcal{L})^{-1/2} \varphi\|_p \approx \|\varphi\|_p$$

by rewriting $\nabla (I + \mathcal{L})^{-1/2}$ as an $L^p(\mu, H)$-valued Riesz transform. For this we need first, the following elementary

**Lemma 1:**   Let $K$ be any function on $[0, 2\pi]$ such that

$$K(\theta) - \tfrac{1}{2} \cot \tfrac{\theta}{2} \in L^\infty([0, \pi]),$$

then the operator $f \to T_K f$ defined by

$$T_K f(x) = \frac{1}{\pi} p.v. \int_0^\pi (f(x + t) - f(x - t)) K(t) dt$$

is again a bounded operator on $L^p([0, 2\pi])$ with

$$\|T_K f\|_p \le B_p \|f\|_p \quad \text{for any } p \in (1, \infty)$$

where $B_p$ depends only on $p$.

**Proof:**   In fact we have

$$|T_K f - \tilde{f}|(x) \le \frac{1}{\pi} \int_0^\pi |f(x + t) - f(x - t)| \, |K(t) - \tfrac{1}{2} \cot \tfrac{\theta}{2}| dt$$

$$\le c \|f\|_{L^p} \|K - \tfrac{1}{2} \cot \tfrac{\theta}{2}\|_{L^\infty}.$$

Hence

$$\|T_K f\|_p \le (c \|K - \tfrac{1}{2} \cot \tfrac{\theta}{2}\|_{L^\infty} + A_p) \|f\|_p.$$

$$\text{QED}$$

**Remark:**   If for some $a \ne 0$, $aK(\theta) - \tfrac{1}{2} \cot \tfrac{\theta}{2} \in L^\infty([0, 2\pi])$, then we have

$$\|T_K f\|_p = \frac{1}{|a|} \|a T_k f\|_p \le \frac{1}{|a|} \|a T_K f - \tilde{f}\|_p + \frac{1}{|a|} \|\tilde{f}\|_p$$

$$\le \frac{1}{|a|} \|a K - \tfrac{1}{2} \cot \tfrac{\theta}{2}\|_{L^\infty} \|f\|_p + \frac{A_p}{|a|} \|f\|_p$$

$$\le c_p \|f\|_p$$

with another constant $c_p$.

**Corollary:** Let $K$ be a function on $[0, \pi]$ such that $K = 0$ on $\left[\frac{\pi}{2}, \pi\right]$ and $K - \frac{1}{2} \cot \frac{\theta}{2} \in L^\infty\left(\left[0, \frac{\pi}{2}\right]\right)$. Then $T_K$ defined by

$$T_K f(x) = \int\limits_0^{\pi/2} (f(x+t) - f(x-t))K(t)dt$$

is continuous from $L^p([0, 2\pi])$ into itself for any $p \in [1, \infty[$.

**Proof:** We have

$$cK(\theta)1_{[0, \frac{\pi}{2}]} - \frac{1}{2} \cot \frac{\theta}{2} \in L^\infty([0, \pi])$$

since on the interval $\left[\frac{\pi}{2}, \pi\right]$, $\sin \frac{\theta}{2} \in \left[\frac{\sqrt{2}}{2}, 1\right]$, then the result follows from the Lemma.                                    QED

## 2   $\nabla(I + \mathcal{L})^{-1/2}$ as the Riesz Transform

Let us denote by $R_\theta(x, y)$ the rotation on $W \times W$ defined by

$$R_\theta(x, y) = (x \cos \theta + y \sin \theta, -x \sin \theta + y \cos \theta).$$

Note that $R_\theta \circ R_\phi = R_{\phi+\theta}$. We have also, putting $e^{-t} = \cos \theta$,

$$
\begin{aligned}
P_t f(x) &= \int\limits_W f(e^{-t}x + \sqrt{1 - e^{-2t}}\, y)\mu(dy) \\
&= \int\limits_W (f \otimes 1)(R_\theta(x, y))\mu(dy) = P_{-\log \cos \theta} f(x).
\end{aligned}
$$

Let us now calculate $(I + \mathcal{L})^{-1/2}\varphi$ using this transformation:

$$
(I + \mathcal{L})^{-1/2}\varphi(x) = \int\limits_0^\infty t^{-1/2}e^{-t}P_t\varphi(x)dt
$$

$$
= \int\limits_0^{\pi/2} (-\log \cos \theta)^{-1/2} \cos \theta \cdot \int\limits_W (\varphi \otimes 1)(R_\theta(x, y))\mu(dy) \tan \theta d\theta
$$

$$
= \int\limits_W \mu(dy)\left[ \int\limits_0^{\pi/2} (-\log \cos \theta)^{-1/2} \sin \theta(\varphi \otimes 1)(R_\theta(x, y))d\theta \right].
$$

On the other hand, we have, for $h \in H$ (even in $C_0^\infty([0, 1])$)

$$\nabla_h P_t \varphi(x)$$

$$= \frac{d}{d\lambda} P_t \varphi(x + \lambda h)|_{\lambda=0}$$

$$= \frac{d}{d\lambda} \int \varphi(e^{-t}(x + \lambda h) + \sqrt{1 - e^{-2t}}\, y)\mu(dy)|_{\lambda=0}$$

$$= \frac{d}{d\lambda} \int \varphi\left(e^{-t}x + \sqrt{1 - e^{-2t}}\left(y + \frac{\lambda e^{-t}}{\sqrt{1 - e^{-2t}}}h\right)\right)\mu(dy)|_{\lambda=0}$$

$$= \frac{d}{d\lambda} \int \varphi(e^{-t}x + \sqrt{1 - e^{-2t}}\, y)\mathcal{E}\left(\frac{\lambda e^{-t}}{\sqrt{1 - e^{-2t}}}I(h)\right)(y)\mu(dy)|_{\lambda=0}$$

$$= \frac{e^{-t}}{\sqrt{1 - e^{-2t}}} \int_W \varphi(e^{-t}x + \sqrt{1 - e^{-2t}}\, y)\delta h(y)\,\mu(dy) .$$

Therefore

$$\nabla_h (I + \mathcal{L})^{-1/2}\varphi(x)$$

$$= \int_0^\infty t^{-1/2} e^{-t} \nabla_h P_t \varphi(x) dt$$

$$= \int_0^\infty t^{-1/2} \frac{e^{-2t}}{\sqrt{1 - e^{-2t}}} \int_W \delta h(y)\varphi(e^{-t}x + \sqrt{1 - e^{-2t}}\, y)\mu(dy)dt$$

$$= \int_0^{\pi/2} (-\log \cos \theta)^{-1/2} \frac{\cos^2 \theta}{\sin \theta} \tan \theta \int \delta h(y)(\varphi \otimes 1)(R_\theta(x, y))\mu(dy)d\theta$$

$$= \int_0^{\pi/2} (-\log \cos \theta)^{-1/2} \cos \theta \int_W \delta h(y) \cdot (\varphi \otimes 1)(R_\theta(x, y))\mu(dy)d\theta$$

Since $\mu(dy)$ is invariant under the transformation $y \mapsto -y$, we have

$$\int \delta h(y)(\varphi \otimes 1)(R_\theta(x, y))\mu(dy) = -\int \delta h(y)(\varphi \otimes 1)(R_{-\theta}(x, y))\mu(dy),$$

therefore:

$$\nabla_h (I + \mathcal{L})^{-1/2} \varphi(x)$$

$$= \int_0^{\pi/2} (-\log \cos \theta)^{-1/2} \cdot$$

$$\int \delta h(y) \frac{(\varphi \otimes 1)(R_\theta(x,y)) - (\varphi \otimes 1)(R_{-\theta}(x,y))}{2} \mu(dy) d\theta$$

$$= \int_W \delta h(y) \int_0^{\pi/2} K(\theta) \left( (\varphi \otimes 1)(R_\theta(x,y)) - (\varphi \otimes 1)(R_{-\theta}(x,y)) \right) d\theta \mu(dy),$$

where $K(\theta) = \frac{1}{2} \cos \theta (-\log \cos \theta)^{-1/2}$.

**Lemma 2:**  We have

$$2K(\theta) - \cot \frac{\theta}{2} \in L^\infty (0, \frac{\pi}{2}]).$$

**Proof:**  The only problem is when $\theta \to 0$. To see this let us put $e^{-t} = \cos \theta$, then

$$\cot \frac{\theta}{2} = \frac{\sqrt{1 + e^{-t}}}{\sqrt{1 - e^{-t}}} \approx \frac{2}{\sqrt{t}}$$

and

$$K(\theta) = \frac{e^{-t}}{\sqrt{t}} \approx \frac{1}{\sqrt{t}}$$

hence

$$2K(\theta) - \cot \frac{\theta}{2} \in L^\infty ([0, \frac{\pi}{2}]).$$

QED

Using Lemma 1, the remark following it and the corollary, we see that the map $f \mapsto p.v. \int_0^{\pi/2} (f(x + \theta) - f(x - \theta)) K(\theta) d\theta$ is a bounded map from $L^p[0, \pi]$ into itself. Moreover

**Lemma 3:**  Let $F : W \times W \to \mathbf{R}$ be a measurable, bounded function. Define $TF(x, y)$ as

$$TF(x, y) = p.v. \int_0^{\pi/2} (F \circ R_\theta(x, y) - F \circ R_{-\theta}(x, y)) K(\theta) d\theta .$$

Then, for any $p > 1$, there exists some $c_p > 0$ such that

$$\|TF\|_{L^p(\mu \times \mu)} \le c_p \|F\|_{L^p(\mu \times \mu)} .$$

**Proof:** We have

$$(TF)(R_\beta(x,y)) = p.v. \int_0^{\pi/2} (F(R_{\beta+\theta}(x,y)) - F(R_{\beta-\theta}(x,y)))K(\theta)d\theta \,,$$

this is the Riesz transform for fixed $(x,y) \in W \times W$, hence we have

$$\int_0^{\pi/2} |TF(R_\beta(x,y))|^p d\beta \le c_p \int_0^\pi |F(R_\beta(x,y))|^p d\beta \,,$$

taking the expectation with respect to $\mu \times \mu$, which is invariant under $R_\beta$, we have

$$\begin{aligned}
E_{\mu\times\mu} \int_0^\pi |TF(R_\beta(x,y))|^p d\beta &= E_{\mu\times\mu} \int_0^\pi |TF(x,y)|^p d\beta \\
&= \frac{\pi}{2} E[|TF|^p] \\
&\le c_p E \int_0^\pi |F(R_\beta(x,y))|^p d\beta \\
&= \pi c_p E[|F|^p] \,.
\end{aligned}$$

QED

We have

**Theorem 1:** $\nabla \circ (I+\mathcal{L})^{-1/2} : L^p(\mu) \to L^p(\mu,H)$ is continuous for any $p > 1$.

**Proof:** With the notations of Lemma 3, we have

$$\nabla_h (I+\mathcal{L})^{-1/2}\varphi = \int_W \delta h(y)\, T(\varphi \otimes 1)(x,y)\mu(dy) \,.$$

From Schwarz inequality:

$$|\nabla(I+\mathcal{L})^{-1/2}\varphi|_H^2 \le \int_W |T(\varphi \otimes 1)(x,y)|^2 \mu(dy)$$

hence, for $p \ge 2$,

$$\begin{aligned}
E[|\nabla(I+\mathcal{L})^{-1/2}\varphi|_H^p] &\le E\left[\left(\int_W |T(\varphi \otimes 1)(x,y)|^2 \mu(dy)\right)^{p/2}\right] \\
&\le E \int_W |T(\varphi \otimes 1)(x,y)|^p \mu(dy) \\
&\le \int\int |(\varphi \otimes 1)(x,y)|^p \mu(dy)\mu(dx) = \|\varphi\|_{L^p(\mu)}^p \,.
\end{aligned}$$

For the case $1 < q < 2$, let $\varphi$ and $\psi$ be smooth (i.e., cylindrical), since $\delta \circ \nabla = \mathcal{L}$, we have, for $p^{-1} + q^{-1} = 1$ (hence $p > 2$ !):

$$E[\varphi\psi] =$$
$$E[(\nabla(I(+\mathcal{L})^{-1/2}\varphi, \nabla(I + \mathcal{L})^{-1/2}\psi)]$$
$$+ \quad E[(I + \mathcal{L})^{-1/2}\varphi.(I + \mathcal{L})^{-1/2}\psi],$$

hence

$$E[(\nabla(I + \mathcal{L})^{-1/2}\varphi, \nabla(I + \mathcal{L})^{-1/2}\psi)_H] = E[\varphi\psi] - E[(I + \mathcal{L})^{-1}\varphi.\psi].$$

Since $(I + \mathcal{L})^{-1}$ is continuous on $L^p(\mu)$ (it is a contraction), we have

$$\sup_{\|\varphi\|_p \leq 1} |E[(\nabla(I + \mathcal{L})^{-1/2}\varphi, \nabla(I + \mathcal{L})^{-1/2}\psi)_H]| \leq c\|\psi\|_q,$$

hence $\|\nabla(I + \mathcal{L})^{-1/2}\psi\|_q \leq \tilde{c}_q\|\psi\|_q$.      QED

**Corollary 1:** We have

$$\|(I + \mathcal{L})^{-1/2}\delta\xi\|_p \leq c_p\|\xi\|_p,$$

for any $\xi \in L^p(\mu; H)$ for $p \in ]1, \infty[$.

**Proof:** Just take the adjoint of $\nabla(I + \mathcal{L})^{-1/2}$.      QED

**Corollary 2:** We have

i) $\|\nabla\varphi\|_p \leq c_p\|(I + \mathcal{L})^{1/2}\varphi\|_p$

ii) $\|(I + \mathcal{L})^{1/2}\varphi\|_p \leq \tilde{c}_p(\|\varphi\|_p + \|\nabla\varphi\|_p)$.

**Proof:**

i) $\|\nabla\varphi\|_p = \|\nabla(I + \mathcal{L})^{-1/2}(I + \mathcal{L})^{1/2}\varphi\|_p \leq c_p\|(I + \mathcal{L})^{1/2}\varphi\|_p$.

ii) $\|(I + \mathcal{L})^{1/2}\varphi\|_p = \|(I + \mathcal{L})^{-1/2}(I + \mathcal{L})\varphi\|_p$

$$= \|(I + \mathcal{L})^{-1/2}(I + \delta\nabla)\varphi\|_p$$
$$\leq \|(I + \mathcal{L})^{-1/2}\varphi\|_p + \|(I + \mathcal{L})^{-1/2}\delta\nabla\varphi\|_p$$
$$\leq \|\varphi\|_p + c_p\|\nabla\varphi\|_p,$$

where the last inequality follows from the Corollary 1.      QED

# Chapter III

# Hypercontractivity

## Hypercontractivity

We know that the semi-group of Ornstein-Uhlenbeck is a bounded operator on $L^p(\mu)$, for any $p \in [1, \infty]$. In fact for $p \in ]1, \infty[$, it is more than bounded. It increases the degree of integrability, this property is called **hypercontractivity** and it is used to show the continuity of linear operators on $L^p(\mu)$-spaces defined via the Wiener chaos decomposition or the spectral decomposition of the Ornstein-Uhlenbeck operator. We shall use it in the next chapter to complete the proof of the Meyer inequalities. Hypercontractivity has been first discovered by E. Nelson, here we follow the proof given by [14].

In the sequel we shall show that this result can be proved using the Ito formula. Let $(\Omega, \mathcal{A}, P)$ be a probability space with $(\mathcal{B}_t; t \in \mathbf{R}_+)$ being a filtration. We take two Brownian motions $(X_t; t \geq 0)$ and $(Y_t; t \geq 0)$ which are not necessarily independent, i.e., $X$ and $Y$ are two continuous, real martingales such that $(X_t^2 - t)$ and $(Y_t^2 - t)$ are again martingales (with respect to $(\mathcal{B}_t)$) and that $X_t - X_s$ and $Y_t - Y_s$ are independent of $\mathcal{B}_s$, for $t > s$. Moreover there exists $(\rho_t; t \in \mathbf{R}_+)$, progressively measurable with values in $[-1, 1]$ such that

$$\left(X_t Y_t - \int_0^t \rho_s \, ds, t \geq 0\right)$$

is again a $(\mathcal{B}_t)$-martingale. Let us denote by

$$\mathcal{X}_t = \sigma(X_s; s \leq t), \quad \mathcal{Y}_t = \sigma(Y_s; s \leq t)$$

i. e., the correponding filtrations of $X$ and $Y$ and by $\mathcal{X}$ and by $\mathcal{Y}$ their respective supremum.

**Lemma 1:** 1) For any $\varphi \in L^1(\Omega, \mathcal{X}, P), t \geq 0$, we have

$$E[\varphi|\mathcal{B}_t] = E[\varphi|\mathcal{X}_t] \text{ a.s.}$$

2)   For any $\psi \in L^1(\Omega, \mathcal{Y}, P)$, $t \geq 0$, we have

$$E[\psi|\mathcal{B}_t] = E[\psi|\mathcal{Y}_t] \text{ a.s.}$$

**Proof:**  1)  From Lévy's theorem, we have also that $(X_t)$ is an $(\mathcal{X}_t)$-Brownian motion. Hence

$$\varphi = E[\varphi] + \int_0^\infty H_s dX_s$$

where $H$ is $(\mathcal{X}_t)$-adapted process. Hence

$$E[\varphi|\mathcal{B}_t] = E[\varphi] + \int_0^t H_s dX_s = E[\varphi|\mathcal{X}_t].$$

$$\text{QED}$$

Let us look at the operator $T : L^1(\Omega, \mathcal{X}, P) \rightarrow L^1(\Omega, \mathcal{Y}, P)$ which is the restriction of $E[\cdot|\mathcal{Y}]$ to the space $L^1(\Omega, \mathcal{X}, P)$. We know that $T : L^p(\mathcal{X}) \rightarrow L^p(\mathcal{Y})$ is a contraction for any $p \geq 1$. In fact, if we impose some conditions to $\rho$, then we have more:

**Proposition 1:**  If $|\rho_t(w)| \leq r$ $(dt \times dP \text{ a.s.})$ for some $r \in [0, 1]$, then $T : L^p(\mathcal{X}) \rightarrow L^q(\mathcal{Y})$ is a bounded operator, where

$$p - 1 \geq r^2(q - 1).$$

**Proof:**  $p = 1$ is already known. So suppose $p, q \in ]1, \infty[$. Since $L^\infty(\mathcal{X})$ is dense in $L^p(\mathcal{X})$, it is enough to prove that $\|TF\|_q \leq \|F\|_p$ for any $F \in L^\infty(\mathcal{X})$. Moreover, since $T$ is a positive operator, we have $|T(F)| \leq T(|F|)$, hence we can work as well with $F \in L_+^\infty(\mathcal{X})$.
From the duality between $L^p$-spaces, we have to show that

$$E[T(F)G] \leq \|F\|_p \|G\|_{q'}, \qquad \left(\frac{1}{q'} + \frac{1}{q} = 1\right),$$

for any $F \in L_+^\infty(\mathcal{X})$, $G \in L_+^\infty(\mathcal{Y})$. Moreover, we can suppose without loss of generality that $F, G \in [a, b]$ a.s. where $0 < a < b < \infty$ (since such random variables are total in all $L_+^p$-spaces, i.e., they separate $L_+^p$ for any $p > 1$).
Let

$$M_t = E[F^p|\mathcal{X}_t]$$
$$N_t = E[G^{q'}|\mathcal{Y}_t].$$

Then, from the Ito representation theorem we have

$$M_t = M_0 + \int_0^t \phi_s dX_s$$

$$N_t = N_0 + \int_0^t \psi_s dY_s$$

where $\phi$ is $\mathcal{X}$-adapted, $\psi$ is $\mathcal{Y}$-adapted, $M_0 = E[F^p]$, $N_0 = E[G^{q'}]$. From the Ito formula, we have

$$M_t^\alpha N_t^\beta = M_0^\alpha N_0^\beta + \int_0^t \alpha M_s^{\alpha-1} N_s^\beta dM_s + \beta \int_0^t M_s^\alpha N_s^{\beta-1} dN_s +$$

$$+ \frac{1}{2} \int_0^t M_s^\alpha N_s^\beta A_s ds$$

where

$$A_t = \alpha(\alpha-1)\left(\frac{\phi_t}{M_t}\right)^2 + 2\alpha\beta \frac{\phi_t}{M_t}\frac{\psi_t}{N_t}\rho_t + \beta(\beta-1)\left(\frac{\psi_t}{N_t}\right)^2$$

and $\alpha = \frac{1}{p}$, $\beta = \frac{1}{q'}$.

[To see this we have

$$M_t^\alpha = M_0^\alpha + \alpha \int_0^t M_s^{\alpha-1}\phi_s dX_s + \frac{\alpha(\alpha-1)}{2} \int_0^t M_s^{\alpha-2}\phi_s^2 ds$$

$$N_t^\beta = \cdots$$

hence

$$M_t^\alpha N_t^\beta - M_0^\alpha N_0^\beta$$

$$= \int_0^t M_s^\alpha dN_s^\beta + \int_0^t N_s^\beta dM_s^\alpha + \alpha\beta \int_0^t M_s^{\alpha-1} N_s^{\beta-1}\phi_s \psi_s \rho_s ds$$

$$= \int_0^t M_s^\alpha \left(\beta N_s^{\beta-1}\psi_s dY_s + \frac{\beta(\beta-1)}{2} N_s^{\beta-2}\psi_s^2 ds\right)$$

$$+ \int_0^t N_s^\beta \left(\alpha M_s^{\alpha-1}\phi_s dX_s + \frac{\alpha(\alpha-1)}{2} M_s^{\alpha-2}\phi_s^2 ds\right)$$

$$+ \alpha\beta \int_0^t M_s^{\alpha-1} N_s^{\beta-1}\phi_s \psi_s \rho_s ds$$

then put together all terms with "$ds$".]

As everything is square integrable, we have

$$
\begin{aligned}
E[M_\infty^\alpha N_\infty^\beta] &= E\Big[E[X^p|\mathcal{X}_\infty]^\alpha \cdot E[Y^{q'}|\mathcal{Y}_\infty]^\beta\Big] \\
&= E[X \cdot Y] \\
&= \frac{1}{2}\int_0^\infty E[N_t^\beta M_t^\alpha A_t]dt + E M_0^\alpha N_0^\beta \\
&= E[X^p]^\alpha E[Y^{q'}]^\beta + \frac{1}{2}\int_0^\infty E[M_t^\alpha N_t^\beta A_t]dt \, .
\end{aligned}
$$

Hence

$$
\boxed{\,E[XY] - \|X\|_p\|Y\|_{q'} = \frac{1}{2}\int_0^\infty E[M_t^\alpha N_t^\beta A_t]dt\,}
$$

Now look at $A_t$ as a polynomial of second degree with respect to $\frac{\phi}{M}$ . Then $\frac{\delta}{4} = \alpha^2\beta^2\rho_t^2 - \alpha(\alpha-1)\beta(\beta-1)$. If $\delta \leq 0$, then the sign of $A_t$ is same as the sign of $\alpha(\alpha-1) \leq 0$, i.e., if $\rho_t^2 \leq \frac{(\alpha-1)(\beta-1)}{\alpha\beta} = \left(1-\frac{1}{\alpha}\right)\left(1-\frac{1}{\beta}\right) = (p-1)(q'-1)$ a.s., then we obtain

$$
E[XY] = E[T(X)Y] \leq \|X\|_p\|Y\|_{q'} \, . \qquad\qquad \text{QED}
$$

**Lemma**   Let $(w,z) = W \times W$ be independent Brownian paths. For $\rho \in [0,1]$, define $x = \rho w + \sqrt{1-\rho^2}\, z$, $\mathcal{X}_\infty$ the $\sigma$-algebra associated to the paths $x$. Then we have

$$
E[F(w)|\mathcal{X}_\infty] = \int_W F(\rho x + \sqrt{1-\pi^2}\, z)\mu(dz).
$$

**Proof:**   For any $G \in L^\infty(\mathcal{X}_\infty)$, we have

$$
\begin{aligned}
E[F(w) \cdot G(x)] &= E[F(w)G(\rho w + \sqrt{1-\rho^2}\, z)] \\
&= E[F(\rho w + \sqrt{1-\rho^2}\, z)G(w)] \\
&= \iint F(\rho\bar{w} + \sqrt{1-\rho^2}\, \bar{z})G(\bar{w}) \cdot \mu(d\bar{w})\mu(d\bar{z}) \\
&= E[G(x)\int F(\rho x + \sqrt{1-\rho^2}\, \bar{z}) \cdot \mu(d\bar{z})]
\end{aligned}
$$

where $\bar{w}, \bar{z}$ represent the dummy variables of integration.          QED

**Corollary 1:**   Under the hypothesis of the above lemma, we have

$$
\left\|\int_W F(\rho x + \sqrt{1-\rho^2}\, \bar{z})\mu(d\bar{z})\right\|_{L^q(d\mu)} \leq \|F\|_{L^p}
$$

for any $(p-1) \geq \rho^2(q-1)$.

# Chapter IV

# $L^p$-Multipliers Theorem, Meyer Inequalities and Distributions

## 1  $L^p$-Multipliers Theorem

$L^p$-Multipliers Theorem gives us a tool to perform some sort of symbolic calculus to study the continuity of the operators defined via the Wiener chaos decomposition of the Wiener functionals. With the help of this calculus we will complete the proof of the Meyer's inequalities.

Almost all of these results have been discovered by P. A. Meyer (cf. [13]) and they are consequences of the Nelson's hypercontractivity theorem.

First let us give first the following simple and important result:

**Theorem 1:**  Let $F \in L^p(\mu)$ and $F = \sum_n I_n(F_n)$ its Wiener chaos development. Then the map $F \rightarrow I_n(F_n)$ is continuous on $L^p(\mu)$.

**Proof:**  Suppose first $p > 2$. Let $t$ be such that $p = e^{2t} + 1$, then we have

$$\|P_t F\|_p \leq \|F\|_2 .$$

Moreover
$$\|P_t I_n(F_n)\|_p \leq \|I_n(F_n)\|_2 \leq \|F\|_2 \leq \|F\|_p$$
but $P_t I_n(F_n) = e^{-nt} I_n(F_n)$, hence

$$\|I_n(F_n)\|_p \leq e^{nt}\|F\|_p .$$

For $1 < p < 2$ we use the duality: let $F \to I_n(F_n) = J_n(F)$. Then

$$
\begin{aligned}
\|I_n(F)\|_p &= \sup_{\|G\|_q \le 1} |\langle G, J_n(F) \rangle| \\
&= \sup |\langle J_n(G), F \rangle| \\
&= \sup |\langle J_n G, J_n F \rangle| \\
&\le \sup e^{nt} \|G\|_q \|F\|_p \\
&= e^{nt} \|F\|_p .
\end{aligned}
$$

QED

**Proposition 1: Multiplier's theorem**

Let the function $h$ be defined as

$$
h(x) = \sum_{k=0}^{\infty} a_k x^k
$$

be an analytic function around the origin with $\sum_k |a_k| \left(\frac{1}{n^\alpha}\right)^k < +\infty$ for $n \ge n_0$, for some $n_0 \in \mathbf{N}$. Let $\phi(x) = h(x^{-\alpha})$ and define $T_\phi$ on $L^p(\mu)$ as

$$
T_\phi F = \sum_{n=0}^{\infty} \phi(n) I_n(F_n) .
$$

Then the operator to $T_\phi$ is bounded on $L^p(\mu)$ for any $p > 1$.

**Proof:**  Suppose first $\alpha = 1$. Let $T_\phi = T_1 + T_2$ where

$$
T_1 F = \sum_{n=0}^{n_0 - 1} \phi(n) I_n(F_n), \qquad T_2 F = (I - T_1) F .
$$

From the hypercontractivity, $F \mapsto T_1 F$ is continuous on $L^p(\mu)$. Let

$$
\Delta_{n_0} F = \sum_{n=n_0}^{\infty} I_n(F_n).
$$

Since

$$
(I - \Delta_{n_0})(F) = \sum_{n=0}^{n_0 - 1} I_n(F_n),
$$

$\Delta_{n_0} : L^p \to L^p$ is continuous, hence $P_t \Delta_{n_0} : L^p \to L^p$ is also continuous. Applying Riesz-Thorin interpolation theorem, which says that if $P_t \Delta_{n_0}$ is $L^q \to L^q$ and $L^2 \to L^2$ then it is $L^p \to L^p$ for any $p$ such that $\frac{1}{p}$ is in the interval $\left[\frac{1}{q}, \frac{1}{2}\right]$, we obtain

$$
\|P_t \Delta_{n_0}\|_{p,p} \le \|P_t \Delta_{n_0}\|_{2,2}^\theta \|P_t \Delta_{n_0}\|_{q,q}^{1-\theta} \le \|P_t \Delta_{n_0}\|_{2,2}^\theta \|\Delta_{n_0}\|_{q,q}^{1-\theta}
$$

where $\frac{1}{p} = \frac{\theta}{2} + \frac{1-\theta}{q}$, $\theta \in ]0,1[$. Choose $q$ large enough such that $\theta \approx 1$ (if necessary). Hence we have

$$\|P_t \Delta_{n_0}\|_{p.p} \le e^{-n_0 t \theta} K, \qquad K = K(n_0, \theta).$$

Similar argument holds for $p \in ]1, 2[$ by duality.

We have

$$
\begin{aligned}
T_2(F) &= \sum_{n \ge n_0} \phi(n) I_n(F_n) = \\
&= \sum_{n \ge n_0} \left( \sum_k a_k \left(\frac{1}{n}\right)^k \right) I_n(F_n) \\
&= \sum_k a_k \sum_{n \ge n_0} \left(\frac{1}{n}\right)^k I_n(F_n) \\
&= \sum_k a_k \sum_{n \ge n_0} \mathcal{L}^{-k} I_n(F_n) \\
&= \sum_k a_k \mathcal{L}^{-k} \Delta_{n_0} F.
\end{aligned}
$$

We also have

$$
\begin{aligned}
\|\mathcal{L}^{-1} \Delta_{n_0} F\|_p &= \left\| \int_0^\infty P_t \Delta_{n_0} F \, dt \right\|_p \le K \int_0^\infty e^{-n_0 \theta t} \|F\|_p \, dt \le K \cdot \frac{\|F\|_p}{n_0 \theta} \\
\|\mathcal{L}^{-2} \Delta_{n_0} F\|_p &= \left\| \int_0^\infty \int_0^\infty P_{t+s} \Delta_{n_0} F \, dt \, ds \right\|_p \le K \cdot \frac{\|F\|_p}{(n_0 \theta)^2}, \\
&\cdots \\
\|\mathcal{L}^{-k} \Delta_{n_0} F\|_p &\le K \|F\|_p \frac{1}{(n_0 \theta)^k}.
\end{aligned}
$$

Therefore

$$\|T_2(F)\|_p \le \sum_k K \|F\|_p \frac{1}{n_0^k \theta^k} \cong \sum K \|F\|_p \frac{1}{n_0^k}$$

by the hypothesis (take $n_0 + 1$ instead of $n_0$ if necessary).

For the case $\alpha \in ]0, 1[$, let $\theta_t^{(\alpha)}(ds)$ be the measure on $\mathbf{R}_+$, defined by

$$\int_{\mathbf{R}_+} e^{-\lambda s} \theta_t^{(\alpha)}(ds) = e^{-t\lambda^\alpha}.$$

Define

$$Q_t^\alpha F = \sum e^{-n^\alpha t} I_n(F_n) = \int_0^\infty P_s F \theta_t^{(\alpha)}(ds).$$

Then

$$\|Q_t^\alpha \Delta_{n_0} F\|_p \ \leq\ \|F\|_p \int_0^\infty e^{-n_0\theta s}\theta_t^{(\alpha)}(ds)$$

$$= \ \|F\|_p e^{-t(n_0\theta)^\alpha},$$

the rest of the proof goes as in the case $\alpha = 1$.                          QED

**Examples of application:**

1) Let

$$\phi(n) \ = \ \left(\frac{1+\sqrt{n}}{\sqrt{1+n}}\right)^s \qquad s \in ]-\infty, \infty[$$

$$= \ h\left(\sqrt{\frac{1}{n}}\right), \qquad h(x) = \left(\frac{1+x}{\sqrt{1+x^2}}\right)^s.$$

Then $T_\phi : L^p \to L^p$ is bounded. Moreover $\phi^{-1}(n) = \frac{1}{\phi(n)} = h^{-1}\left(\sqrt{\frac{1}{n}}\right)$, $h^{-1}(x) = \frac{1}{h(x)}$ is also analytic near the origin, hence $T_{\phi^{-1}} : L^p \to L^p$ is also a bounded operator.

2) Let $\phi(n) = \frac{\sqrt{1+n}}{\sqrt{2+n}}$ then $h(x) = \sqrt{\frac{x+1}{2x+1}}$ satisfies also the above hypothesis.

3) As an application of (2), look at

$$\begin{aligned}
\|(I + \mathcal{L})^{1/2}\nabla\varphi\|_p &= \ \|\nabla(2I + \mathcal{L})^{1/2}\varphi\|_p \\
&\leq \ \|(I + \mathcal{L})^{1/2}(2I + \mathcal{L})^{1/2}\varphi\|_p \\
&= \ \|(2I + \mathcal{L})^{1/2}(I + \mathcal{L})^{1/2}\varphi\|_p \\
&= \ \|T_\phi(I + \mathcal{L})^{1/2}(I + \mathcal{L})^{1/2}\varphi\|_p \\
&\leq \ c_p\|(I + \mathcal{L})\varphi\|_p.
\end{aligned}$$

Continuing this way we can show that

$$\begin{aligned}
\|\nabla^k\varphi\|_{L^p(\mu, H^{\otimes k})} &\leq \ c_{p,k}\|\varphi\|_{p,k}(= \|(I + \mathcal{L})^{k/2}\varphi\|_p) \\
&\leq \ \tilde{c}_{p,k}(\|\varphi\|_p + \|\nabla^k\varphi\|_{L^p(\mu, H^{\otimes k})})
\end{aligned}$$

and this completes the proof of the Meyer inequalities for the scalar-valued Wiener functionals. If $\mathcal{X}$ is a separable Hilbert space, we denote with $D_{p,k}(\mathcal{X})$ the completion of the $\mathcal{X}$-valued polynomials with respect to the norm

$$\|\alpha\|_{D_{p,k}(\mathcal{X})} = \|(I + \mathcal{L})^{k/2}\|_{L^p(\mu, \mathcal{X})}.$$

We define as in the case $\mathcal{X} = \mathbf{R}$, the Sobolev derivative $\nabla$, the divergence $\delta$, etc. All we have said for the real case extend trivially to the vector case, including the Meyer inequalities. In fact, in the proof of these inequalities the main step is the Riesz inequality for the Hilbert transform. However this inequality is also true for any Hilbert space (in fact it holds also for a class of Banach spaces which contains Hilbert spaces, called UMD spaces). The rest is almost the transcription of the real case combined with the Khintchine inequalities. We leave hence this passage to the reader. QED

**Corollary** For every $p > 1$, $k \in \mathbf{R}$, $\nabla$ has a continuous extension as a map $D_{p,k} \to D_{p,k-1}(H)$.

**Proof:** We have

$$
\begin{aligned}
\|\nabla \varphi\|_{p,k} &= \|(I + \mathcal{L})^{k/2} \nabla \varphi\|_p \\
&= \|\nabla (2I + \mathcal{L})^{k/2} \varphi\|_p \\
&\leq c_p \|(1 + \mathcal{L})^{1/2}(2I + \mathcal{L})^{k/2} \varphi\|_p \\
&\approx \|(I + \mathcal{L})^{(k+1)/2} \varphi\|_p \\
&= \|\varphi\|_{p,k+1} .
\end{aligned}
$$

QED

**Corollary** $\delta = \nabla^* : D_{p,k}(H) \to D_{p,k-1}$ is continuous $\forall p > 1$ and $k \in \mathbf{R}$.

**Proof:** The proof follows from the duality. QED

In particular:

**Corollary**

- $$
\nabla : \bigcap_{p,k} D_{p,k} = D \to D(H) = \bigcap_{p,k} D_{p,k}(H)
$$

is continuous and extends continuously as a map

$$
\nabla : D' = \bigcup_{p,k} D_{p,k} \to D'(H) = \bigcup_{p,k} D_{p,k}(H),
$$

the elements of the space $D'$ are called Meyer-Watanabe distributions.

- $$
\delta : \bigcap_{p,k} D_{p,k}(H) = D(H) \to D
$$

is continuous and has a continuous extension

$$
\delta : D'(H) \to D'
$$

**Proof:** Everything follows from the dualities

$$(D)' = D', \ (D(H))' = D'(H).$$

QED

**Definition:** For $n \geq 1$, we define $\delta^n$ as $(\nabla^n)^*$ with respect to $\mu$.

**Proposition 2** For $\varphi \in L^2(\mu)$, we have

$$\varphi = E[\varphi] + \sum_{n \geq 1} \frac{1}{n!} \delta^n(E[\nabla^n \varphi]).$$

**Proof:** First suppose that $h \mapsto \varphi(w + h)$ is analytic for almost all $w$. Then we have

$$\varphi(w + h) = \varphi(w) + \sum_{n \geq 1} \frac{(\nabla^n \varphi(w), h^{\otimes n})_{H^{\otimes n}}}{n!}.$$

Take the expectations:

$$
\begin{aligned}
E[\varphi(w + h)] &= E[\varphi \cdot \mathcal{E}(\delta h)] \\
&= E[\varphi] + \sum_{n} \frac{(E[\nabla^n \varphi], h^{\otimes n})}{n!} \\
&= E[\varphi] + \sum_{n \geq 1} E\left[ \frac{I_n(E[\nabla^n \varphi])}{n!} \mathcal{E}(\delta h) \right].
\end{aligned}
$$

Since the finite linear combinations of the elements of the set $\{\mathcal{E}(\delta h); h \in H\}$ is dense in any $L^p(\mu)$, we obtain the identity

$$\varphi(w) = E[\varphi] + \sum_{n \geq 1} \frac{I_n(E[\nabla^n \varphi])}{n!}.$$

Let $\psi \in D$, then we have (with $E[\psi] = 0$),

$$
\begin{aligned}
\langle \varphi, \psi \rangle &= \sum_{n \geq 1} E[I_n(\varphi_n) I_n(\psi_n)] = \\
&= \sum_{n} E\left[ \frac{I_n(E[\nabla^n \varphi])}{n!} \cdot I_n(\psi_n) \right] = \\
&= \sum_{n} (E[\nabla^n \varphi], \psi_n) = \\
&= \sum_{n} \frac{1}{n!} (E[\nabla^n \varphi], E[\nabla^n \psi]) \\
&= \sum_{n} \frac{1}{n!} E[(E[\nabla^n \varphi], \nabla^n \psi)] \\
&= \sum_{n} \frac{1}{n!} E[\delta^n(E[\nabla^n \varphi]) \cdot \psi]
\end{aligned}
$$

hence we obtain that

$$\varphi = \sum_n \frac{1}{n!} \delta^n E[\nabla^n \varphi],$$

in particular $\delta^n E[\nabla^n \varphi] = I_n(E[\nabla^n \varphi])$.                    QED

Let us give another result important for the applications:

**Proposition 3** Let $F$ be in some $L^p(\mu)$ with $p > 1$ and suppose that the distributional derivative $\nabla F$ of $F$, is in some $L^r(\mu, H)$, $(1 < r)$. Then $F$ belongs to $D_{r \wedge p, 1}$.

**Proof:** Without loss of generality, we can assume that $r \leq p$. Let $(e_i; i \in \mathbf{N})$ be a complete, orthonormal basis of the Cameron-Martin space $H$. Denote by $V_n$ the sigma-field generated by $\delta e_1, \ldots, \delta e_n$, and by $\pi_n$ the ortohogonal projection of $H$ onto the subspace spanned by $e_1, \ldots, e_n$, $n \in \mathbf{N}$. Let us define $F_n$ by

$$F_n = E[P_{1/n} F | V_n],$$

where $P_{1/n}$ is the Ornstein-Uhlenbeck semigroup at $t = 1/n$. Then $F_n$ belongs to $D_{r,k}$ for any $k \in \mathbf{N}$ and converges to $F$ in $L^r(\mu)$. Moreover, from Doob's lemma, $F_n$ is of the form

$$F_n(w) = \alpha(\delta e_1, \ldots, \delta e_n),$$

with $\alpha$ being a Borel function on $\mathbf{R}^n$, which is in the intersection of the Sobolev spaces $\cap_k W_{r,k}(\mathbf{R}^n, \mu_n)$ defined with the Ornstein-Uhlenbeck operator $L_n = -\Delta + x \cdot \nabla$ on $\mathbf{R}^n$. Since $L_n$ is elliptic, the Weyl lemma implies that $\alpha$ can be chosen as a $C^\infty$-function. Consequently, $\nabla F_n$ is again $V_n$-measurable and we find , using the very definition of conditional expectation and the Mehler formula, that

$$\nabla F_n = E[e^{-1/n} \pi_n P_{1/n} \nabla F | V_n].$$

Consequently, from the martingale convergence theorem and from the fact that $\pi_n \to I_H$ in the operators' norm topology, it follows that

$$\nabla F_n \to \nabla F,$$

in $L^r(\mu, H)$, consequently $F$ belongs to $D_{r,1}$.                    Q.E.D.

## Appendix: Passing from the classical Wiener space to the AWS (or vice-versa):

Let $(W, H, \mu)$ be an abstract Wiener space. Since, à priori, there is no notion of time, it seems that we can not define the notion of anticipation, non-anticipation, etc.

**This difficulty can be overcome in the following way:**

Let $(p_\lambda; \lambda \in \Sigma)$, $\Sigma \subset \mathbf{R}$, be a resolution of identity on the separable Hilbert space $H$, i.e., each $p_\lambda$ is an orthogonal projection, increasing to $I_H$, in the sense that $\lambda \mapsto (p_\lambda h, h)$ is an increasing function. Let us denote by $H_\lambda = \overline{p_\lambda(H)}$

**Definition 1:** We will denote by $\mathcal{F}_\lambda$ the $\sigma$-algebra generated by the real polynomials $\varphi$ on $W$ such that $\nabla\varphi \in H_\lambda$ $\mu$-almost surely.

**Lemma 1:** We have

$$\bigvee_{\lambda\in\Sigma} \mathcal{F}_\lambda = \mathcal{B}(W)$$

up to $\mu$-negligeable sets.

**Proof:** We have already $\bigvee \mathcal{F}_\lambda \subset \mathcal{B}(W)$. Conversely, if $h \in H$, then $\nabla\delta h = h$. Since $\bigcup_{\lambda\in\Sigma} H_\lambda$ is dense in $H$, there exists $(h_n) \subset \bigcup_\lambda H_\lambda$ such that $h_n \to h$ in $H$. Hence $\delta h_n \to \delta h$ in $L^p(\mu)$, $\forall p \geq 1$. Since each $\delta h_n$ is $\bigvee \mathcal{F}_\lambda$-measurable, so does $\delta h$. Since $\mathcal{B}(W)$ is generated by $\{\delta h; h \in H\}$ the proof is completed.          QED

**Definition 2:** A random variable $\xi : W \to H$ is called a simple, adapted vector field if it can be written as

$$\xi = \sum_{i<+\infty} F_i(p_{\lambda_{i+1}}h_i - p_{\lambda_i}h_i)$$

where $h_i \in H$, $F_i$ are $\mathcal{F}_{\lambda_i}$-measurable (and smooth for the time being!) random variables.

**Proposition 4** For each adapted simple vector field we have

i) $\delta\xi = \sum_{i<+\infty} F_i\delta(p_{\lambda_{i+1}}h_i - p_{\lambda_i}h_i)$

ii) $E[(\delta\xi)^2] = E[|\xi|_H^2]$.

**Proof:** i) We have

$$\delta[F_i(p_{\lambda_{i+1}} - p_{\lambda_i})h_i] = F_i\delta[(p_{\lambda_{i+1}} - p_{\lambda_i})h_i] - (\nabla F_i, (p_{\lambda_{i+1}} - p_{\lambda_i})h_i)\,.$$

Since $\nabla F_i \in H_\lambda$, the second term is null.

(ii) is well-known.                                                    QED

**Remark:** If we note $\sum F_i\, 1_{]\lambda_i,\lambda_{i+1}]}(\lambda)h_i$ by $\dot{\xi}(\lambda)$, we have the following notations:

$$\delta\xi = \delta\int_\Sigma \dot{\xi}(\lambda)dp_\lambda \quad \text{with} \quad \|\delta\xi\|_2^2 = E\int_\Sigma d(\dot{\xi}_\lambda, p_\lambda\dot{\xi}_\lambda) = \|\xi\|_{L^2(\mu,H)}^2\,,$$

which are significantly analogous to the things that we have seen before as the Ito stochastic integral.

Now the Ito representation theorem holds in this setting also: suppose $(p_\lambda; \lambda \in \Sigma)$ is continuous, then:

**Theorem 2** Let us denote with $D^a_{2,0}(H)$ the completion of adapted simple vector fields with respect to the $L^2(\mu, H)$-norm. Then we have

$$L_2(\mu) = \mathbf{R} + \{\delta\xi : \xi \in D^a_{2,0}(H)\},$$

i.e., any $\varphi \in L_2(\mu)$ can be written as

$$\varphi = E[\varphi] + \delta\xi$$

for some $\xi \in D^a_{2,0}(H)$. Moreover such $\xi$ is unique up to $L^2(\mu, H)$-equivalence classes.

The following result explains the reason of the existence of the Brownian motion (cf. also [26]):

**Theorem 3** Suppose that there exists some $\Omega_0 \in H$ such that the set $\{p_\lambda \Omega_0; \lambda \in \Sigma\}$ has a dense span in $H$ (i.e. the linear combinations from it is a dense set). Then the real-valued $(\mathcal{F}_\lambda)$-martingale defined by

$$b_\lambda = \delta p_\lambda \Omega_0$$

is a Brownian motion with a deterministic time change and $(\mathcal{F}_\lambda; \lambda \in \Sigma)$ is its canonical filtration completed with the negligeable sets.

**Example:** Let $H = H_1([0, 1])$, define $A$ as the operator defined by $Ah(t) = \int_0^t s\dot{h}(s)ds$. Then $A$ is a self-adjoint operator on $H$ with a continuous spectrum which is equal to $[0, 1]$. Moreover we have

$$(p_\lambda h)(t) = \int_0^t 1_{[0,\lambda]}(s)\dot{h}(s)ds$$

and $\Omega_0(t) = \int_0^t 1_{[0,1]}(s)ds$ satisfies the hypothesis of the above theorem. $\Omega_0$ is called the vacuum vector (in physics).

This is the main example, since all the (separable) Hilbert spaces are isomorphic, we can carry this time structure to any abstract Hilbert-Wiener space as long as we do not need any particular structure of time.

# Chapter V

# Some applications of the distributions

## Introduction

In this chapter we give some applications of the extended versions of the derivative and the divergence operators. First we give an extension of the Ito-Clark formula to the space of the scalar distributions. We refer the reader to [2] and [17] for the developments of this formula in the case of Sobolev differentiable Wiener functionals . Let us briefly explain the problem: although, we know from the Ito representation theorem, that each square integrable Wiener functional can be represented as the stochastic integral of an adapted process, without the use of the distributions, we can not calculate this process, since any square integrable random variable is not neccessarily in $D_{2,1}$, hence it is not Sobolev differentiable in the ordinary sense. As it will explained, this problem is completely solved using the differentiation in the sense of distributions. Afterwards we give a straightforward application of this result to prove a $0 - 1$ law for the Wiener measure. At the second section we construct the composition of the tempered distributions with nondegenerate Wiener functionals as Meyer-Watanabe distributions. This construction carries also the information that the probability density of a nondegenerate random variable is not only infinitely differentiable but also it is rapidly decreasing. The same idea is then applied to prove the regularity of the solutions of the Zakai equation for the filtering of non-linear diffusions.

# 1   Extension of the Ito-Clark formula

Let $F$ be any integrable random variable. Then we know that $F$ can be represented as

$$F = E[F] + \int_0^1 H_s dW_s ,$$

where $(H_s; s \in [0, 1])$ is an adapted process such that, it is unique and

$$\int_0^1 H_s^2 ds < +\infty \quad \text{a.s.}$$

Moreover, if $F \in L^p$ $(p > 1)$, then we also have

$$E\left[\left(\int_0^1 |H_s|^2 ds\right)^{p/2}\right] < +\infty.$$

One question is how to calculate the process $H$. In fact, below we will extend the Ito representation and answer to the above question for any $F \in D'$ (i.e., the Meyer-Watanabe distributions).

We begin with:

**Lemma 1**  Let $\xi \in D(H)$, then $\pi\xi$ defined by $\pi\xi(t) = \int_0^t E[\dot{\xi}_s | \mathcal{F}_s] ds$ belongs again to $D(H)$, i.e. $\pi : D(H) \to D(H)$ is continuous.

**Proof:**   We have $\mathcal{L}\pi\xi = \pi\mathcal{L}\xi$, hence

$$
\begin{aligned}
\|\pi\xi\|_{p,k} &= E\left[\left(\int_0^1 |(I + \mathcal{L})^{k/2} E[\dot{\xi}_s | \mathcal{F}_s]|^2 ds\right)^{p/2}\right] = \\
&= E\left[\left(\int_0^1 |E[(I + \mathcal{L})^{k/2} \dot{\xi}_s | \mathcal{F}_s]|^2 ds\right)^{p/2}\right] \\
&\leq c_p E\left[\left(\int_0^1 |(I + \mathcal{L}^{k/2} \dot{\xi}_s|^2 ds\right)^{p/2}\right] \quad (c_p \cong p)
\end{aligned}
$$

where the last inequality follows from the convexity inequalities of the dual predictable projections (c.f. Dellacherie-Meyer, Vol. 2).                                  QED

**Lemma 2:**   $\pi : D(H) \to D(H)$ extends as a continuous mapping to $D'(H) \to D'(H)$.

**Proof:** Let $\xi \in D(H)$, then we have, for $k > 0$,

$$
\begin{aligned}
\|\pi\xi\|_{p,-k} &= \|(I + \mathcal{L})^{-k/2}\pi\xi\|_p \\
&= \|\pi(I + \mathcal{L})^{-k/2}\xi\|_p \leq c_p\|(I + \mathcal{L})^{-k/2}\xi\|_p \\
&\leq c_p\|\xi\|_{p,-k},
\end{aligned}
$$

then the proof follows since $D(H)$ is dense in $D'(H)$.

QED

Before going further let us give a notation: if $F$ is in some $D_{p,1}$ then its Gross-Sobolev derivative $\nabla F$ gives an $H$-valued random variable. Hence $t \mapsto \nabla F(t)$ is absolutely continuous with respect to the Lebesgue measure on $[0, 1]$, we denote by $D_s F$ its Radon-Nikodym derivative with respect to the Lebesgue measure. Note that it is $ds \times d\mu$-almost everywhere well-defined.

**Lemma 3:** Let $\varphi \in D$, then we have

$$
\begin{aligned}
\varphi &= E[\varphi] + \int_0^1 E[D_s\varphi|\mathcal{F}_s]dW_s \\
&= E[\varphi] + \delta\pi\nabla\varphi.
\end{aligned}
$$

Moreover $\pi\nabla\varphi \in D(H)$.

**Proof:** Let $U$ be an element of $L^2(\mu, H)$ such that $u(t) = \int_0^t \dot{u}_s ds$ with $(\dot{u}_t; t \in [0, 1])$ being an adapted and bounded process. Then we have, from the Girsanov theorem,

$$
E[\varphi(w + \lambda u(w)). \exp(-\lambda \int_0^1 \dot{u}_s dW_s - \frac{\lambda^2}{2} \int_0^1 \dot{u}_s ds)] = E[\varphi].
$$

Differentiating both sides at $\lambda = 0$, we obtain:

$$
E[(\nabla\varphi(w), u) - \varphi \int_0^1 \dot{u}_s dW_s] = 0,
$$

i.e.,

$$
E[(\nabla\varphi, u)] = E[\varphi \int_0^1 \dot{u}_s dW_s].
$$

Furthermore

$$E\left[\int_0^1 D_s\varphi \dot{u}_s ds\right] \;=\; E\left[\int_0^1 E[D_s\varphi|\mathcal{F}_s]\dot{u}_s ds\right]$$

$$=\; E[(\pi\nabla\varphi, u)_H]$$

$$=\; E\left[\left(\int_0^1 E[D_s\varphi|\mathcal{F}_s]dW_s\right)\left(\int_0^1 \dot{u}_s dW_s\right)\right].$$

Since the set $\{\int_0^1 \dot{u}_s dW_s, \dot{u}$ as above$\}$ is dense in $L_0^2(\mu) = L^2(\mu) - \langle L^2(\mu), 1\rangle$, we see that

$$\varphi - E[\varphi] = \int_0^1 E[D_s\varphi|\mathcal{F}_s]dW_s = \delta\pi\nabla\varphi \,.$$

The rest is obvious from the Lemma 1.                                         QED

**Theorem 1:**   For any $T \in D'$, we have

$$T = \langle T, 1\rangle + \delta\pi\nabla T \,.$$

**Proof:**   Let $(\varphi_n) \subset D$ such that $\varphi_n \to T$ in $D'$. Then we have

$$T \;=\; \lim_n \varphi_n$$

$$=\; \lim_n[E[\varphi_n] + \delta\pi\nabla\varphi_n]$$

$$=\; \lim_n E[\varphi_n] + \lim_n \delta\pi\nabla\varphi_n$$

$$=\; \lim_n\langle 1, \varphi_n\rangle + \lim_n \delta\pi\nabla\varphi_n$$

$$=\; \langle 1, T\rangle + \delta\pi\nabla T$$

since $\nabla : D' \to D'(H)$, $\pi : D'(H) \to D'(H)$ and $\delta : D'(H) \to D'$ are all linear, continuous mappings.                                         QED

Here is a nontrivial application of the Theorem 1:

**Theorem 2:** (0–1 law)   Let $A \in \mathcal{B}(W)$ such that $A + H = A$. Then $\mu(A) = 0$ or 1.

**Proof:**   $A + H = A$ implies that

$$1_A(w + \lambda h) = 1_A(w) \text{ a.s.}$$

hence $\nabla 1_A = 0$. Consequently, Theorem 1 implies that

$$1_A = \langle 1_A, 1\rangle = \mu(A) \Rightarrow \mu(A)^2 = \mu(A).$$                                         QED

# 2 Lifting of $\mathcal{S}'(\mathbf{R}^d)$ with random variables

Let $f : \mathbf{R} \to \mathbf{R}$ be a $C_b^1$-function, $F \in D$. Then we know that

$$\nabla(f(F)) = f'(F)\nabla F.$$

Now suppose that $|\nabla F|_H^{-2} \in \bigcap L^p(\mu)$, then

$$f'(F) = \frac{(\nabla(f(F)), \nabla F)_H}{|\nabla F|_H^2}.$$

Even if $f$ is not $C^1$, the right hand side of this equality has a sense if we look at $\nabla(f(F))$ as an element of $D'$. In the following we will develop this idea:

**Definition:** Let $F : W \to \mathbf{R}^d$ be a random variable such that $F_i \in D$, $\forall i = 1, \dots, d$, and that

$$[\det(\nabla F_i, \nabla F_j)]^{-1} \in \bigcap_{p>1} L^p(\mu).$$

Then we say that $F$ is a **non-degenerate** random variable.

**Lemma 1** Let us denote by $\sigma_{ij} = (\nabla F_i, \nabla F_j)$ and by $\gamma = \sigma^{-1}$ (as a matrix). Then $\gamma \in D(\mathbf{R}^d \otimes \mathbf{R}^d)$.

**Proof:** Formally, we have, using the relation $\sigma \cdot \gamma = Id$,

$$\nabla \gamma_{ij} = \sum_{k,l} \gamma_{ik}\gamma_{jl}\nabla\sigma_{kl}.$$

To justify this we define first $\sigma_{ij}^\epsilon = \sigma_{ij} + \epsilon\delta_{ij}$, $\epsilon > 0$. Then we can write $\gamma_{ij}^\epsilon = f_{ij}(\sigma^\epsilon)$, where $f : \mathbf{R}^d \otimes \mathbf{R}^d \to \mathbf{R}^d \otimes \mathbf{R}^d$ is a smooth function of polynomial growth. Hence $\gamma_{ij}^\epsilon \in D$. Then from the dominated convergence theorem we have $\gamma_{ij}^\epsilon \to \gamma_{ij}$ in $L^p$ as well as $\nabla^k\gamma_{ij}^\epsilon \xrightarrow{L^p} \nabla^l\gamma_{ij}$ (this follows again from $\gamma^\epsilon \cdot \sigma^\epsilon = Id$). QED

**Lemma 2** Let $G \in D$. Then we have, $\forall f \in \mathcal{S}(\mathbf{R}^d)$

i) $E[\partial_i f(F).G] = E[f(F).l_i(G)]$

where $G \mapsto l_i(G)$ is linear and for any $1 < r < g < \infty$,

$$\sup_{\|G\|_{q,1} \leq 1} \|l_i(G)\|_r < +\infty.$$

ii) Similarly

$$E[\partial_{i_1 \dots i_k} f \circ F.G] = E[f(F) \cdot l_{i_1 \dots i_k}(G)]$$

and

$$\sup_{\|G\|_{q,1} \leq 1} \|l_{i_1 \dots i_k}(G)\|_r < \infty.$$

**Proof:**   We have

$$\nabla(f \circ F) = \Sigma \partial_i f(F) \nabla F_i \Rightarrow (\nabla(f \circ F), \nabla F_j) = \Sigma \sigma_{ij} \partial_i f(F).$$

Since $\sigma$ is invertible, we obtain:

$$\partial_i f(F) = \sum_j \gamma_{ij}(\nabla(f \circ F), \nabla F_j).$$

Then

$$
\begin{aligned}
E[\partial_i f(F).G] &= \sum_j E[\gamma_{ij}(\nabla(f \circ F), \nabla F_j).G] \\
&= \sum_j E[f \circ F.\delta\{\nabla F_j \gamma_{ij} G\}],
\end{aligned}
$$

hence we see that $l_i(G) = \sum_j \delta\{\nabla F_j \gamma_{ij} G\}$. We have

$$
\begin{aligned}
l_i(G) &= -\sum_j [(\nabla(\gamma_{ij} G), \nabla F_j) - \gamma_{ij} G \mathcal{L} F_j] \\
&= -\sum_j [\gamma_{ij}(\nabla G, \nabla F_j) - \sum_{k,l} \gamma_{ik} \gamma_{jl}(\nabla \sigma_{kl}, \nabla F_j)G - \gamma_{ij} G \mathcal{L} F_j].
\end{aligned}
$$

Hence

$$
|l_{ij}(G)| \le \sum_j \Big[ \sum_{kl} |\gamma_{ik}\gamma_{jl}||\nabla\sigma_{kl}||\nabla F_j||G| \;+\; |\gamma_{ij}||\nabla F_j||\nabla G| +
$$
$$
+\; |\gamma_{ij}||G||\mathcal{L} F_j| \Big].
$$

Choose $p$ such that $\frac{1}{r} = \frac{1}{p} + \frac{1}{q}$ and apply Hölder's inequality:

$$
\begin{aligned}
\|l_i(G)\|_r &\le \sum_{j=1}^d \Big[ \sum_{k,l} \|G\|_q \|\gamma_{ik}\gamma_{jl}|\nabla\sigma_{kl}|_H|\nabla F_j|_H\|_p + \\
&\qquad + \|\gamma_{ij}|\nabla F_j|\|_p \||\nabla G|\|_q + \|\gamma_{ij}\mathcal{L} F_j\|_p \|G\|_q \Big] \\
&\le \|G\|_{q,1} \Big[ \sum_{j=1}^d \|\gamma_{ik}\gamma_{jl}|\nabla F_{kl}||\nabla F_j|\|_p + \\
&\qquad + \|\gamma_{ij}|\nabla F_j|\|_p + \|\gamma_{ij}\mathcal{L} F_j\|_p \Big].
\end{aligned}
$$

ii) For $i > 1$ we iterate this procedure.                          QED

   **Now remember that** $\mathcal{S}(\mathbb{R}^d)$ can be written as the intersection (i.e., projective limit) of the following Banach spaces:
   Let $A = I - \Delta + |x|^2$, $\|f\|_{2k} = \|A^k f\|_\infty$ (the uniform norm) and $S_{2k} = $ completion of $\mathcal{S}(\mathbb{R}^d)$ with respect to the norm $\|\cdot\|_{2k}$.

**Theorem 1** Let $F \in D(\mathbf{R}^d)$ be a non-degenerate random variable. Then we have for $f \in \mathcal{S}(\mathbf{R}^d)$:

$$\|f \circ F\|_{p,-2k} \le c_{p,k} \|f\|_{-2k} \, .$$

**Proof:** Let $\psi = A^{-k} f \in \mathcal{S}(\mathbf{R}^d)$. For $G \in D$, we know that there exists some $\eta_{2k}(G) \in D$ ($G \mapsto \eta_{2k}(G)$ is linear) from the Lemma 2, such that

$$E[A^k \psi \circ F.G] = E[\psi \circ F.\eta_{2k}(G)] \, ,$$

i.e.,

$$E[f \circ F.G] = E[(A^{-k} f)(F).\eta_{2k}(G)] \, .$$

Hence

$$|E[f \circ F.G]| \le \|A^{-k} f\|_\infty \|\eta_{2k}(G)\|_{L^1}$$

and

$$\sup_{\|G\|_{q,2k} \le 1} |E[f \circ F.G]| \le \|A^{-k} f\|_\infty \sup_{\|G\|_{q,2k} \le 1} \|\eta_{2k}(G)\|_1$$
$$= K\|f\|_{-2k} \, .$$

Hence $\|f \circ F\|_{p,-2k} \le K\|f\|_{-2k}$ . $\hspace{2cm}$ QED

**Corollary 1:** The map $f \mapsto f \circ F$ from $\mathcal{S}(\mathbf{R}^d) \to D$ has a continuous extension to $\mathcal{S}'(\mathbf{R}^d) \to D'$.

# Some applications

If $F : W \to \mathbf{R}^d$ is a non-degenerate random variable, then we have seen that the map $f \mapsto f \circ F$ from $\mathcal{S}(\mathbf{R}^d) \to D$ has a continuous extension to $\mathcal{S}'(\mathbf{R}^d) \to D'$, denoted by $T \mapsto T \circ F$.

For $f \in \mathcal{S}(\mathbf{R}^d)$, let us look at the following Pettis integral:

$$\int_{\mathbf{R}^d} f(x) \mathcal{E}_x dx,$$

where $\mathcal{E}_x$ denotes the Dirac measure at $x \in \mathbf{R}^d$. We have, for any $g \in \mathcal{S}(\mathbf{R}^d)$,

$$\left\langle \int f(x)\mathcal{E}_x dx, g \right\rangle = \int \langle f(x)\mathcal{E}_x, g \rangle dx$$
$$= \int f(x)\langle \mathcal{E}_x, g \rangle dx$$
$$= \int f(x)g(x)dx = \langle f, g \rangle \, .$$

Hence we have proven:

**Lemma 1:** The following representation holds in $\mathcal{S}(\mathbf{R}^d)$:

$$f = \int_{\mathbf{R}^d} f(x)\mathcal{E}_x\,dx.$$

From Lemma 1 we have

**Lemma 2:** We have

$$\int \langle \mathcal{E}_y(F), \varphi \rangle f(y)dy = E[f(F).\varphi],$$

for any $\varphi \in D$, where $\langle \cdot, \cdot \rangle$ denotes the bilinear form of duality between $D'$ and $D$.

**Proof:** Let $\rho_\epsilon$ be a mollifier. Then $\mathcal{E}_y * \rho_\epsilon \to \mathcal{E}_y$ in $S'$ on the other hand

$$\int (\mathcal{E}_y * \rho_\epsilon)(F)f(y)dy = \int \rho_\epsilon(F+y).f(y)dy =$$

$$= \int \rho_\epsilon(y)f(y+F)dy \xrightarrow[\epsilon \to 0]{} f(F).$$

On the other hand, for $\varphi \in D$,

$$\lim_{\epsilon \to 0} \int < (\mathcal{E}_y * \rho_\epsilon)(F), \varphi > f(y)dy = \int \lim_{\epsilon \to 0} < (\mathcal{E}_y * \rho_\epsilon)(F), \varphi > f(y)dy$$

$$= \int < \mathcal{E}_y(F), \varphi > f(y)dy$$

$$= < f(F), \varphi >$$

$$= E[f(F)\varphi].$$
                                                                                                                              QED

**Corollary:** We have

$$\langle \mathcal{E}_x(F), 1 \rangle = \frac{d(F^*\mu)}{dx}(x) = p_F(x),$$

moreover $p_F \in \mathcal{S}(\mathbf{R}^d)$ (i.e., the probability density of $F$ is not only $C^\infty$ but it is also a rapidly decreasing function).

**Proof:** We know that the map $T \to E[T(F).\varphi]$ is continuous on $S'(\mathbf{R}^d)$ hence there exists some $p_{F,\varphi} \in \mathcal{S}(\mathbf{R}^d)$ such that

$$E[T(F).\varphi] = {}_S\langle P_{F,\varphi}, T \rangle_{S'}.$$

Let $p_{F,1} = p_F$, then it follows from the Lemma 2 that

$$E[f(F)] = \int \langle \mathcal{E}_y(F), 1 \rangle f(y)dy.$$
                                                                                                                              QED

**Remark:** From the disintegration of measures, we have

$$\int E[\varphi|F=x]f(x)dx = E[f(F)\cdot\varphi]$$
$$= \int f(x)\langle\mathcal{E}_x(F),\varphi\rangle dx$$

hence

$$E[\varphi|F=x] = \langle\mathcal{E}_x(F),\varphi\rangle$$

$dx$-almost surely. In fact the right hand side is an everywhere defined version of this conditional probability.

**Remark:** Let $(x_t)$ be the solution of the following stochastic differential equation:

$$dx_t(w) = b_i(x_t(w))dt + \sigma_i(x_t(w))dw_t^i$$

$$x_0 = x \text{ given},$$

where $b : \mathbf{R}^d \to \mathbf{R}^d$ and $\sigma_i : \mathbf{R}^d \to \mathbf{R}^d$ are smooth vector fields with bounded derivatives. Let us denote by

$$X_0 = \sum_{i=1}^d \tilde{b}_0^i \frac{\partial}{\partial x_i}, \qquad X_j = \sum \sigma_i^j \frac{\partial}{\partial x_j}$$

where

$$\tilde{b}^i(x) = b^i(x) - \frac{1}{2}\sum_{k,\alpha}\partial_k\sigma_\alpha^i(x)\sigma_\alpha^k(x).$$

If the Lie algebra of vector fields generated by $\{X_0, X_1, \ldots, X_d\}$ has dimension equal to $d$ at any $x \in \mathbf{R}^d$, then $x_t(w)$ is non-degenerate cf. [32]. In fact it is also uniformly non-degenerate in the following sense:

$$E\int_s^t |Det(\nabla x_r^i, \nabla x_r^j)|^{-p}dr < \infty, \qquad \forall 0 < s < t, \forall p > 1.$$

As a corollary of this result, combined with the lifting of $\mathcal{S}'$ to $D'$, we can show the following:

For any $T \in \mathcal{S}'(\mathbf{R}^d)$, one has the following:

$$T(x_t) - T(x_s) = \int_s^t AT(x_s)ds + \int_s^t \sigma_{ij}(x_s)\cdot\partial_j T(x_s)dW_s^i,$$

where the Lebesgue integral is a Bochner integral, the stochastic integral is as defined at the first section of this chapter and we have used the following notation:

$$A = \sum b^i\partial_i + \frac{1}{2}\sum a_{ij}(x)\frac{\partial^2}{\partial x_i\partial x_j}, \qquad a(x) = (\sigma\sigma^*)_{ij}, \ \sigma = [\sigma_1, \ldots, \sigma_d].$$

## Applications to the filtering of the diffusions

Suppose that $y_t = \int\limits_0^t h(x_s)ds + B_t$ where $h \in C_b^\infty(\mathbb{R}^d) \otimes \mathbb{R}^d$, $B$ is another Brownian motion independent of $w$ above. $(y_t; t \in [0,1])$ is called an (noisy) observation of $(x_t)$. Let $\mathcal{Y}_t = \sigma\{y_s; s \in [0,t]\}$ be the observed data till $t$. The filtering problem consists of calculating the random measure $f \mapsto E[f(x_t)|\mathcal{Y}_t]$. Let $P^0$ be the probability defined by

$$dP^0 = Z_1^{-1}dP$$

where $Z_t = \exp \int\limits_0^1 h(x_s).dy_s - \frac{1}{2}\int\limits_0^t |h(x_s)|^2 ds$. Then for any bounded, $\mathcal{Y}_t$-measurable random variable $Y_t$, we have:

$$
\begin{aligned}
E[f(x_t).Y_t] &= E\left[\frac{Z_t}{Z_t}f(x_t).Y_t\right] = E^0[Z_t f(x_t)Y_t] \\
&= E^0[E[Z_t f(x_t)|\mathcal{Y}_t] \cdot Y_t] = \\
&= E\left[\frac{1}{E^0[Z_t|\mathcal{Y}_t]}E^0[Z_t f(x_t)|\mathcal{Y}_t] \cdot Y_t\right],
\end{aligned}
$$

hence

$$E[f(x_t)|\mathcal{Y}_t] = \frac{E^0[Z_t f(x_t)|\mathcal{Y}_t]}{E^0\left[Z_t|\mathcal{Y}_t\right]} .$$

If we want to study the smoothness of the measure $f \mapsto E[f(x_t)|\mathcal{Y}_t]$, then from the above formula, we see that it is sufficient to study the smoothness of $f \mapsto E^0[Z_t f(x_t)|\mathcal{Y}_t]$. The reason for the use of $P^0$ is that $w$ and $(y_t; t \in [0,1])$ are two independent Brownian motions under $P^0$ (this follows directly from Paul Lévy's theorem of the characterization of the Brownian motion).

After this preliminaries, we can prove the following

**Theorem**  Suppose that the map $f \mapsto f(x_t)$ from $\mathcal{S}(\mathbb{R}^d)$ into $D$ has a continuous extension as a map from $\mathcal{S}'(\mathbb{R}^d)$ into $D'$. Then the measure $f \mapsto E[f(x_t)|\mathcal{Y}_t]$ has a density in $\mathcal{S}(\mathbb{R}^d)$.

**Proof:**  As explained above, it is sufficient to prove that the (random) measure $f \mapsto E^0[Z_t f(x_t)|\mathcal{Y}_t]$ has a density in $\mathcal{S}(\mathbb{R}^d)$.

Let $\mathcal{L}_y$ be the Ornstein-Uhlenbeck operator on the space of the Brownian motion $(y_t; t \in [0,1])$. Then we have

$$\mathcal{L}_y Z_t = Z_t\left(-\int\limits_0^t h(x_s)dy_s + \frac{1}{2}\int\limits_0^t |h(x_s)|^2 ds\right) \in \bigcap_p L^p$$

It is also easy to see that

$$\mathcal{L}_w^k Z_t \in \bigcap_p L^p .$$

- Hence $Z_t(w, y) \in D(w, y)$, where $D(w, y)$ denotes the space of test functions defined on the product Wiener space with respect to the laws of w and y.

- The second point is that the operator $E^0[\bullet|\mathcal{Y}_t]$ is a continuous mapping from $D_{p,k}^k(w, y)$ into $D_{p,k}^0(y)$ since $\mathcal{L}_y$ commutes with $E^0[\bullet|\mathcal{Y}_t]$ (for any $p \geq 1, k \in \mathbf{Z}$).

- Hence the map
$$T \mapsto E^0[T(x_t)Z_t|\mathcal{Y}_t]$$
is continuous from $\mathcal{S}'(\mathbf{R}^d) \to D'(y)$. In particular, for fixed $T \in \mathcal{S}'$, $\exists p > 1$ and $k \in \mathbf{N}$ such that $T(x_t) \in D_{p,-k}(w)$. Since $Z_t \in D(w, y)$,
$$Z_t T(x_t) \in D_{p,-k}(w, y)$$
and
$$T(x_t).(I + \mathcal{L}_y)^{k/2}Z_t \in D_{p,-k}(w, y).$$
Hence
$$E^0[T(x_t) \cdot (I + \mathcal{L}_y)^{k/2}Z_t|\mathcal{Y}_t] \in D_{p,-k}(y).$$

- Hence
$$(I + \mathcal{L})^{-k/2}E^0[T(x_t)(I + \mathcal{L}_y)^{k/2}Z_t|\mathcal{Y}_t] = E^0[T(x_t)Z_t|\mathcal{Y}_t]$$
belongs to $L^p(y)$. Therefore we see that:
$$T \mapsto E^0[T(x_t)Z_t|\mathcal{Y}_t]$$

defines a linear, continuous (use the closed graph theorem for instance) map from $\mathcal{S}'(\mathbf{R}^d)$ into $L^p(y)$. Since $\mathcal{S}'(\mathbf{R}^d)$ is a nuclear space, the map $T \xrightarrow{\Theta} E^0[T(x_t)Z_t|\mathcal{Y}_t]$ is a nuclear operator, hence by definition it has a representation:
$$\Theta = \sum_{i=1}^{\infty} \lambda_i f_i \otimes \alpha_i$$
where $(\lambda_i) \in l^1$, $(f_i) \subset \mathcal{S}(\mathbf{R}^d)$ and $(\alpha_i) \subset L^p(y)$ are bounded sequences. Define
$$k_t(x, y) = \sum_i \lambda_i f_i(x)\alpha_i(y) \in \mathcal{S}(\mathbf{R}^d)\widetilde{\otimes}_1 L^p(y)$$

where $\widetilde{\otimes}_1$ denotes the projective tensor product topology. It is easy now to see that, for $g \in \mathcal{S}(\mathbf{R}^d)$
$$\int_{\mathbf{R}^d} g(x)k_t(x, y)dx = E^0[g(x_t) \cdot Z_t|\mathcal{Y}_t].$$

QED

# Chapter VI

# Positive distributions and applications

## Positive Meyer-Watanabe distributions

If $\theta$ is a positive distribution on $\mathbf{R}^d$, then a well-known theorem says that $\theta$ is a positive measure, finite on the compact sets. We will prove an analogous result for the Meyer-Watanabe distributions in this section, show that they are absolutely continuous with respect to the capacities defined with respect to the scale of the Sobolev spaces on the Wiener space and give an application to the construction of the local time of the Wiener process. We end the chapter by making some remarks about the Sobolev spaces constructed by the second quantization of an elliptic operator on the Cameron-Martin space.

We will work on the classical Wiener space $C_0([0,1]) = W$. First we have the following:

**Proposition:** Suppose $(T_n) \subset D'$ and each $T_n$ is also a probability on $W$. If $T_n \to T$ in $D'$, then $T$ is also a probability and $T_n \to T$ in the weak topology of measures (on $W$).

**Proof:** It is sufficient to prove that the set of probability measures $(\nu_n)$ associated to $(T_n)$, is tight. In fact, let $S = D \cap C_b(W)$. If the tightness holds, then we will have, for $\nu = w - \lim \nu_n$,

$$\nu(\varphi) = T(\varphi) \quad \text{on } S.$$

Since $S$ is weakly dense in $C_b(W)$ the proof will be completed (remember $e^{i\langle w, w^* \rangle}, w^* \in W^*$, belongs to $S!$).

Let $G(w)$ be defined as

$$G(w) = \int\limits_0^1 \int\limits_0^1 \frac{|w(t) - w(s)|^8}{|t-s|^3} \, ds \, dt.$$

Then, it is not difficult to show that $G \in D$ and $A_\lambda = \{G(w) \le \lambda\}$ is a compact subset of $W$ (cf.[1]).

Moreover, we have $\bigcup\limits_{\lambda \ge 0} A_\lambda = W$ almost surely. Let $\varphi \in C^\infty(\mathbf{R})$ such that $0 \le \varphi \le 1$; $\varphi(x) = 1$ for $x \ge 0$, $\varphi(x) = 0$ for $x \le -1$. Let $\varphi_\lambda(x) = \varphi(x - \lambda)$.

We have

$$\nu_n(A_\lambda^c) \le \int_W \varphi_\lambda(G(w)) \, \nu_n(dw).$$

We claim that

$$\int_W \varphi_\lambda(G) d\nu_n = \langle \varphi_\lambda(G), T_n \rangle.$$

To see this, for $\varepsilon > 0$, write

$$G_\varepsilon(w) = \int\limits_{[0,1]^2} \frac{|w(t) - w(s)|^8}{(\varepsilon + |t-s|)^3} \, ds \, dt.$$

Then $\varphi_\lambda(G_\varepsilon) \in S$ (but not $\varphi_\lambda(G)$, since $G$ is not continuous on $W$!) Since $\varphi_\lambda(G_\varepsilon) \in S = G_b(W) \cap D$, we have

$$\int \varphi_\lambda(G_\varepsilon) d\nu_n = \langle \varphi_\lambda(G_\varepsilon), T_n \rangle.$$

But $\varphi_\lambda(G_\varepsilon) \to \varphi_\lambda(G)$ in $D$, hence

$$\lim_{\varepsilon \downarrow 0} \langle \varphi_\lambda(G_\varepsilon, T_n \rangle = \langle \varphi_\lambda(G), T_n \rangle.$$

From the dominated convergence theorem, we have also

$$\lim_{\varepsilon \to 0} \int \varphi_\lambda(G_\varepsilon) d\nu_n = \int \varphi_\lambda(G) d\nu_n.$$

This proves our claim. Now, since $T_n \to T$ in $D'$, exists some $k > 0$ and $p > 1$ such that $T_n \to T$ in $D_{p,-k}$. Therefore

$$\begin{aligned}\langle \varphi_\lambda(G), T_n \rangle &= \langle (I + \mathcal{L})^{k/2} \varphi_\lambda(G), (I + \mathcal{L})^{-k/2} T_n \rangle \\ &\le \|(I + \mathcal{L})^{k/2} \varphi_\lambda(G)\|_q \cdot \sup_n \|(I + \mathcal{L})^{-k/2} T_n\|_p.\end{aligned}$$

From the Meyer inequalities, we see that

$$\lim_{\lambda \to \infty} \|(I + \mathcal{L})^{k/2} \varphi_\lambda(G)\|_q = 0,$$

in fact, it is sufficient to see that $\nabla^i(\varphi_\lambda(G)) \to 0$ in $L^p$ for all $i \leq [k] + 1$, but this is obvious from the choice of $\varphi_\lambda$.

Therefore we have proven that

$$\lim_{\lambda \to \infty} \sup_n \mu_n(A_\lambda^c) \leq \sup_n \|(I + \mathcal{L})^{-k} T_n\|_p \lim_{\lambda \to \infty} \|(I + \mathcal{L})^k \varphi_\lambda(G)\|_p = 0,$$

which is the definition of tightness. QED

**Corollary:** Let $T \in D'$ such that $\langle T, \varphi \rangle \geq 0$, for all positive $\varphi \in D$. Then $T$ is a Radon measure on $W$.

**Proof:** Let $(h_i) \subset H$ be a complete, orthonormal basis of $H$. Let $V_n = \sigma\{\delta h_1, \ldots, \delta h_n\}$. Define $T_n$ as $T_n = E[P_{1/n} T | V_n]$ where $P_{1/n}$ is the Ornstein-Uhlenbeck semi-group on $W$. Then $T_n \geq 0$ and it is a random variable in some $L^p(\mu)$. Therefore it defines a measure on $W$ (even absolutely continuous with respect to $\mu$!). Moreover $T_n \to T$ in $D'$, hence the proof follows from the proposition. QED

# 1 Capacities and positive Wiener functionals

If $p \in [1, \infty[$, $O \subset W$ is an open set and $k > 0$, we define the $(p, k)$-capacity of $O$ as

- $C_{p,k}(O) = \inf\{\|\varphi\|_{p,k}^p : \varphi \in D_{p,k}, \varphi \geq 1 \ \mu\text{-a.e. on } O\}$.

- • If $A \subset W$ is any subset, define its $(p, k)$-capacity as

$$C_{p,k}(A) = \inf\{C_{p,k}(O); O \text{ is open } O \supset A\}.$$

- We say that some property takes place $(p, k)$-quasi everywhere if the set on which it does not hold has $(p, k)$-capacity zero.

- We say $N$ is a slim set if $C_{p,k}(N) = 0$, $\forall p > 1$, $k > 0$.

- A function is called $(p, k)$-*quasi continuous* if $\forall \varepsilon > 0$, $\exists$ open set $O_\varepsilon$ such that $C_{p,k}(O_\varepsilon) < \varepsilon$ and the function is continuous on $O_\varepsilon^c$.

- It is called $\infty$-quasi continuous if it is $(p, k)$-quasi continuous $\forall(p, k)$. The

following results are proved by Fukushima & Kanako:

**Lemma 1:**

i) If $F \in D_{p,k}$, then there exists a $(p, k)$-quasi continuous function $\tilde{F}$ such that $F = \tilde{F}$ $\mu$-a.e. and $\tilde{F}$ is $(p, k)$-quasi everywhere defined, i.e. if $\tilde{G}$ is another such function, then $C_{p,k}(\{\tilde{F} \neq \tilde{G}\}) = 0$.

ii) If $A \subset W$ is arbitrary, then

$$C_{p,k}(A) = \inf\{\|\varphi\|_{p,k} : \varphi \in D_{p,k}, \quad \check{\varphi} \geq 1 \ (p,r) - q.e. \text{ on } A\}$$

iii) There exists a unique element $U_A \in D_{p,k}$ such that $\tilde{U}_A \geq 1 \ (p,k)$-quasi everywhere on $A$ with $C_{p,k}(A) = \|U_A\|_{p,k}$, and $\tilde{U}_A \geq 0 \ (p,k)$-quasi everywhere. $U_A$ is called the $(p,k)$-equilibrium potential of $A$.

**Theorem 1:**   Let $T \in D'$ be a positive distribution and suppose that $T \in D_{q,-k}$ for some $q > 1$, $k \geq 0$. Then, if we denote by $\nu_T$ the measure associated to $T$, we have

$$\bar{\nu}_T(A) \leq \|T\|_{q,-k}(C_{p,k}(A))^{1/p},$$

for any set $A \subset W$, where $\bar{\nu}_T$ denotes the outer measure with respect to $\nu_T$. In particular $\nu_T$ does not charge the slim sets.

**Proof:**   Let $V$ be an open set in $W$ and let $U_V$ be its equilibrium potential of order $(p,k)$. We have

$$\begin{aligned}
\langle P_{1/n}T, U_V \rangle &= \int P_{1/n}T \, U_V \, d\mu \\
&\geq \int_V P_{1/n}T \, U_V \, d\mu \\
&\geq \int_V P_{1/n}T \, d\mu \\
&= \nu_{P_{1/n}T}(V).
\end{aligned}$$

Since $V$ is open, we have, from the fact that $\nu_{P_{1/n}T} \to \nu_T$ weakly,

$$\liminf_{n \to \infty} \nu_{P_{1/n}T}(V) \geq \nu_T(V).$$

On the other hand

$$\begin{aligned}
\lim_{n \to \infty} \langle P_{1/n}T, U_V \rangle &= \langle T, U_V \rangle \\
&\leq \|T\|_{q,-k}\|U_V\|_{p,k} \\
&= \|T\|_{q,-k}C_{p,k}(V)^{1/p}.
\end{aligned}$$

QED

**An application**

1) Let $f : \mathbf{R}^d \to \mathbf{R}$ be a function from $\mathcal{S}'(\mathbf{R}^d)$ and suppose that $(X_t)$ is a hypoelliptic diffusion on $\mathbf{R}^d$. Then $X_t$ is a nondegenerate random variable in

the sense of the second section of the fifth chapter (cf.[32]). Consequently we have the extension of the Ito formula

$$f(X_t) - f(X_u) = \int_u^t Lf(X_s)ds + \int_u^t \sigma_{ij}(X_s)\partial_i f(X_s)dW_s^j ,$$

with the obvious notations. Note that, since we did not make any differentiability hypothesis about $f$, the above integrals are to be interpreted as the elements of $D'$. Suppose that $Lf$ is a bounded measure on $\mathbf{R}^d$, from our result about the positive distributions, we see that $\int_u^t Lf(X_s)ds$ is a measure on $W$ which does not charge the slim sets. By difference, so does the term $\int_u^t \sigma_{ij}(X_s)\partial_i f(X_s)dW_s^j$.

2) Apply this to $d = 1$, $L = \frac{1}{2}\Delta$ (i.e. $\sigma = 1$), $f(x) = |x|$. Then we have

$$|W_t| - |W_u| = \frac{1}{2}\int_u^t \Delta|x|(W_s)ds + \int_u^t \frac{d}{dx}|x|(W_s)dW_s .$$

As $\frac{d}{dx}|x| = \text{sign}(x)$, we have

$$\int_u^t \frac{d}{dx}|x|(W_s)dW_s = \int_u^t \text{sign}(W_s)dW_s = M_t^u$$

is a measure absolutely continuous with respect to $\mu$. Since $\lim_{u \to 0} M_t^u = N_t$ exists in all $L^p$, so does

$$\lim_{u \to 0} \int_u^t \Delta|x|(W_s)ds$$

in $L^p$ for any $p \geq 1$. Consequently $\int_0^t \Delta|x|(W_s)ds$ is absolutely continuous with respect to $\mu$, i.e., it is a random variable. It is easy to see that

$$\Delta|x|(W_s) = 2\mathcal{E}_0(W_s)$$

i.e., we obtain

$$\int_0^t 2\mathcal{E}_0(W_s)ds \;=\; \int_0^t \Delta|x|(W_s)ds$$
$$=\; 2l_t^0$$

which is the local time of Tanaka. Note that, although $\mathcal{E}_0(W_s)$ is singular with respect to $\mu$, its Pettis integral is absolutely continuous with respect to $\mu$.

2) If $F : W \to \mathbf{R}^d$ is a non-degenerate random variable, then for any $S \in \mathcal{S}'(\mathbf{R}^d)$ with $S \geq 0$ on $\mathcal{S}_+(\mathbf{R}^d)$, $S(F) \in D'$ is a positive distribution, hence it is a positive Radon measure on $W$. In particular $\mathcal{E}_x(F)$ is a positive Radon measure.

# Distributions associated to $\Gamma(A)$

For a "tentative" generality we suppose that $(W, H, \mu)$ is an abstract Wiener space. Let $A$ be a selfadjoint operator on $H$, we suppose that its spectrum lies in $]1, \infty[$, hence $A^{-1}$ is bounded and $\|A^{-1}\| < 1$. Let

$$H_\infty = \bigcap_n Dom(A^n),$$

hence $H_\infty$ is dense in $H$ and $\alpha \mapsto (A^\alpha h, h)$ is increasing. Denote by $H_\alpha$ the completion of $H_\infty$ with respect to the norm $|h|_\alpha^2 = (A^\alpha h, h); \ \alpha \in \mathbf{R}$. Evidently $H_\alpha' \cong H_{-\alpha}$ (isomorphism). If $\varphi : W \to \mathbf{R}$ is a nice Wiener functional with $\varphi = \sum_{n=0}^\infty I_n(\varphi_n)$, define the second quantization of $A$

$$\Gamma(A)\varphi = E[\varphi] + \sum_{n=1}^\infty I_n(A^{\otimes n}\varphi_n).$$

**Definition:** For $p > 1$, $k \in \mathbf{Z}$, $\alpha \in \mathbf{R}$, we define $D_{p,k}^\alpha$ as the completion of polynomials (based on $H_\infty$) with respect to the norm:

$$\|\varphi\|_{p,k;\alpha} = \|(I + \mathcal{L})^{k/2}\Gamma(A^{1/2})\varphi\|_{L^p(\mu)},$$

where $\varphi(w) = \text{polynomial}(\delta h_1, \ldots, \delta h_n), \ h_i \in H_\infty$.

If $\mathcal{X}$ is a separable Hilbert space, $D_{p,k}^\alpha(\mathcal{X})$ is defined likewise.

**Remark:** i) If $\varphi = \exp(\delta h - \frac{1}{2}|h|^2)$ then we have

$$\Gamma(A)\varphi = \exp \delta(Ah) - \frac{1}{2}|Ah|^2.$$

ii) $D_{p,k}^\alpha$ is decreasing with respect to $\alpha, p$ and $k$.

**Theorem 1:** Let $(W^\alpha, H_\alpha, \mu_\alpha)$ be the abstract Wiener space corresponding to the Cameron-Martin space $H_\alpha$. Let us denote by $D_{p,k}^{(\alpha)}$ the Sobolev space on $W^\alpha$ defined by

$$\|\varphi\|_{D_{p,k}^{(\alpha)}} = \|(I + \mathcal{L})^{k/2}\varphi\|_{L^p(\mu_\alpha, W^\alpha)}$$

Then $D_{p,k}^{(\alpha)}$ and $D_{p,k}^\alpha$ are isomorphic.

**Remark:** This isomorphism is not algebraic, i.e., it does not commute with the pointwise multiplication.

**Proof:** We have

$$E[e^{i\delta(A^{\alpha/2}h)}] = \exp \frac{1}{2}|A^{\alpha/2}h|^2 = \exp \frac{|h|_\alpha^2}{2}$$

which is the characteristic function of $\mu_\alpha$ on $W^\alpha$.                    QED

**Theorem 2:** i) For $p > 2$, $\alpha \in \mathbf{R}$, $k \in \mathbf{Z}$, there exists some $\beta > \frac{\alpha}{2}$ such that

$$\|\varphi\|_{D_{p,k}^\alpha} \leq \|\varphi\|_{D_{2,k}^\beta}$$

consequently $\bigcap\limits_{\alpha,k} D_{2,k}^\alpha = \bigcap\limits_{\alpha,p,k} D_{p,k}^\alpha$ .

ii) Moreover, for some $\beta > \alpha$ we have

$$\|\varphi\|_{D_{p,k}^\alpha} \leq \|\varphi\|_{D_{2,0}^\beta} .$$

**Proof:** i) We have

$$\begin{aligned}
\|\varphi\|_{D_{p,k}^\alpha} &= \left\| \sum_n (1+n)^{k/2} I_n((A^{\alpha/2})^{\otimes n} \varphi_n) \right\|_{L^p} \\
&= \left\| \sum (1+n)^{k/2} e^{nt} e^{-nt} I_n((A^{\alpha/2})^{\otimes n} \varphi_n) \right\|_{L^p} .
\end{aligned}$$

From the hypercontractivity of $P_t$, we can choose $t$ such that $p = e^{2t} + 1$ then

$$\left\| \sum (1+n)^{k/2} e^{nt} e^{-nt} I_n(\ldots) \right\|_p \leq \left\| \sum (1+n)^{k/2} e^{nt} I_n(\ldots) \right\|_2 .$$

Choose $\beta > 0$ such that $\|A^{-\beta}\| \leq e^{-t}$, hence

$$\begin{aligned}
&\left\| \sum (1+n)^{k/2} e^{nt} I_n(\ldots) \right\|_2 \\
\leq\ & \left\| \sum (1+n)^{k/2} \Gamma(A^\beta) \Gamma(A^{-\beta}) e^{nt} I_n((A^{\alpha/2})^{\otimes n} \varphi_n) \right\|_2 \\
\leq\ & \sum (1+n)^{k/2} \| I_n((A^{\beta+\alpha/2})^{\otimes n} \varphi_n)) \|_2 \\
=\ & \|\varphi\|_{D_{2,k}^{2\beta+\alpha}} .
\end{aligned}$$

ii) If we choose $\|A^{-\beta}\| < e^{-t}$ then the difference suffices to absorb the action of the multiplicator $(1+n)^{k/2}$ which is of polynomial growth and the former gives an exponential decrease. **QED**

**Corollary 1:** We have similar relations for the any separable Hilbert space valued functionals.

**Proof:** Use the Khintchine inequality.

As another corollary we have

**Corollary 2:** i) $\nabla : \Phi \to \Phi(H_\infty) = \bigcap\limits_\alpha \Phi(H_\alpha)$ and $\delta : \Phi(H_\infty) \to \Phi$ are continuous. Consequently $\nabla$ and $\delta$ have continuous extensions as linear operators $\Phi' \to \Phi'(H_{-\infty})$ and $\Phi'(H_{-\infty}) \to \Phi'$.

ii) $\Phi$ is an algebra.

iii) For any $T \in \Phi'$, there exists some $\zeta \in \Phi'(H_{-\infty})$ such that $T = \langle T, 1 \rangle + \delta \zeta$.

**Proof:**  i) Follows from Theorem 1 and 2.

ii) It is sufficient to show that $\varphi^2 \in \Phi$ if $\varphi \in \Phi$. This follows from the multiplication formula of the multiple Wiener integrals. (left to the reader).

iii) If $T \in \Phi'$, then there exists some $\alpha > 0$ such that $T \in D_{2,0}^{-\alpha}$, i.e., $T$ under the isomorphism of Theorem 1 is in $L^2(\mu_\alpha, W^\alpha)$ on which we have Ito representation (cf. Appendix to the Chapter IV).

**Proposition:**  Suppose that $A^{-1}$ is $p$-nuclear, i.e., $\exists p \geq 1$ such that $A^{-p}$ is nuclear. Then $\Phi$ is a nuclear Fréchet space.

**Proof:**  This goes as in the white noise case, except that the eigenvectors of $\Gamma(A^{-1})$ are of the form $H_{\vec{a}}(\delta h_{\alpha_1}, \ldots, \delta h_n)$ with $h_{\alpha_i}$ are the eigenvectors of $A$. QED

# Applications to positive distributions

Let $T \in \Phi'$ be a positive distribution. Then there exists some $D_{p,-k}^{-\alpha}$ such that $T \in D_{p,-k}^{-\alpha}$ and $\langle T, \varphi \rangle \geq 0$ for any $\varphi \in D_{q,k}^{\alpha}$, $\varphi \geq 0$. Hence $i_\alpha(T)$ is a positive functional on $D_{1,k}^{(\alpha)}$ (i.e., the Sobolev space on $W^\alpha$). Therefore $i_\alpha(T)$ is a Radon measure on $W^{-\alpha}$. Hence we find that, in fact the support of $T$ is $W^{-\alpha}$ which is much smaller than $H_{-\infty}$.

**Open question:**  Is there a smallest $W^{-\alpha}$? If yes, can one characterize it?

# Chapter VII

# Characterization of independence of some Wiener functionals

## 1 Independence of Wiener functionals

In probability theory, one of the most important and difficult properties is the independence of random variables. In fact, even in the elementary probability, the tests required to verify the independence of three or more random variables get very quickly quite difficult. Hence it is very tempting to try to characterize the independence of random variables via the local operators that we have seen in the preceeding chapters.

Let us begin with two random variables: let $F, G \in D_{p,1}$ for some $p > 1$. They are independent if and only if

$$E[e^{i\alpha F}e^{i\beta G}] = E[e^{i\alpha F}]E[e^{i\beta G}]$$

for any $\alpha, \beta \in \mathbf{R}$, which is equivalent to

$$E[a(F)b(G)] = E[a(F)]E[b(G)]$$

for any $a, b \in C_b(\mathbf{R})$.

Let us denote by $\tilde{a}(F) = a(F) - E[a(F)]$, then we have: $F$ and $G$ are independent if and only if

$$E[\tilde{a}(F) \cdot b(G)] = 0, \qquad \forall a, b \in C_b(\mathbf{R}).$$

Since $e^{i\alpha x}$ can be approximated pointwise with smooth functions, we can suppose as well that $a, b \in C_b^1(\mathbf{R})$ (or $C_0^\infty(\mathbf{R})$). Since $\mathcal{L}$ is invertible on the centered random variables, we have

$$
\begin{aligned}
E[\tilde{a}(F)b(G)] &= E[\mathcal{L}\mathcal{L}^{-1}\tilde{a}(F) \cdot b(G)] \\
&= E[\delta \nabla \mathcal{L}^{-1}\tilde{a}(F) \cdot b(G)] \\
&= E[(\nabla \mathcal{L}^{-1}\tilde{a}(F), \nabla(b(G)))_H] \\
&= E[((I+\mathcal{L})^{-1}\nabla a(F), \nabla(b(G)))] \\
&= E[((I+\mathcal{L})^{-1}(a'(F)\nabla F), b'(G)\nabla G)_H] \\
&= E[b'(G) \cdot ((I+\mathcal{L})^{-1}(a'(F)\nabla F), \nabla G)_H] \\
&= E[b'(G) \cdot E[((I+\mathcal{L})^{-1}(a'(F)\nabla F, \nabla G)_H|\sigma(G)]] \,.
\end{aligned}
$$

In particular choosing $a = e^{i\alpha x}$, we find that

**Proposition 1:**   $F$ and $G$ (in $D_{p,1}$) are independent if and only if

$$
E[((I+\mathcal{L})^{-1}(e^{i\alpha F}\nabla F), \nabla G)_H|\sigma(G)] = 0 \quad \text{a.s.}
$$

However this result is not very useful, this is because of the non-localness property of the operator $\mathcal{L}^{-1}$. Let us however look at the case of multiple Wiener integrals:

First recall the following multiplication formula of the multiple Wiener integrals:

**Lemma 1:**   Let $f \in \hat{L}^2([0,1]^p)$, $g \in \hat{L}^2([0,1]^q)$. Then we have

$$
I_p(f) \cdot I_q(q) = \sum_{m=0}^{p \wedge q} \frac{p!q!}{m!(p-m)!(q-m)!} I_{p+q-2m}(f \otimes_m g) \,,
$$

where $f \otimes_m g$ denotes the contraction of order $m$ of the tensor $f \otimes g$ (i.e., the partial scalar product of $f$ and $g$ in $L^2[0,1]^m$).

By the help of this lemma we will prove:

**Theorem 1:**   $I_p(f)$ and $I_q(g)$ are independent if and only if

$$
f \otimes_1 g = 0 \quad \text{a.s. on } [0,1]^{p+q-2} \,.
$$

**Proof:**   $(\Rightarrow)$ : By independence, we have

$$
E[I_p^2 I_q^2] = p!\|f\|^2 q!\|g\|^2 = p!q!\|f \otimes g\|^2 \,.
$$

On the other hand

$$
I_p(f)I_q(g) = \sum_0^{p \wedge q} m! C_p^m C_q^m \, I_{p+q-2m}(f \otimes_m g) \,,
$$

hence

$$E[(I_p(f)I_q(g))^2]$$

$$= \sum_0^{p \wedge q} (m! C_p^m C_q^m)^2 (p+q-2m)! \|f \hat{\otimes}_m g\|^2$$

$$\geq (p+q)! \|f \hat{\otimes} g\|^2 \quad \text{(dropping the terms with } m \geq 1) .$$

We have, by definition:

$$\|f \hat{\otimes} g\|^2 = \left\| \frac{1}{(p+q)!} \sum_{\sigma \in S_{p+q}} f(t_{\sigma(1)}, \dots, t_{\sigma(p)}) g(t_{\sigma(p+1)}, \dots, t_{\sigma(p+q)}) \right\|^2$$

$$= \frac{1}{((p+q)!)^2} \sum_{\sigma, \pi \in S_{p+q}} \lambda_{\sigma, \pi} ,$$

where $S_{p+q}$ denotes the group of permutations of order $p+q$ and

$$\lambda_{\sigma, \pi} = \int_{[0,1]^{p+q}} f(t_{\sigma(1)}, \dots, t_{\sigma(p)}) g(t_{\sigma(p+1)}, \dots, t_{\sigma(p+q)}) \cdot$$
$$f(t_{\pi(1)}, \dots, t_{\pi(p)}) g(t_{\pi(p+1)}, \dots, t_{\pi(p+q)}) dt_1 \dots dt_{p+q} .$$

Without loss of generality, we may suppose that $p \leq q$. Suppose now that $(\sigma(1), \dots, \sigma(p))$ and $(\pi(1), \dots, \pi(p))$ has $k \geq 0$ elements in common. If we use the block notations, then

$$(t_{\sigma(1)}, \dots, t_{\sigma(p)}) = (A_k, \tilde{A})$$
$$(t_{\sigma(p+1)}, \dots, t_{\sigma(p+q)}) = B$$
$$(t_{\pi(1)}, \dots, t_{\pi(p)}) = (A_k, \tilde{C})$$
$$(t_{\pi(p+1)}, \dots, t_{\pi(p+q)}) = D$$

where $A_k$ is the subblock containing elements common to $(t_{\pi(1)}, \dots, t_{\pi(p)})$ and $(t_{\sigma(1)}, \dots, t_{\sigma(p)})$. Then we have

$$\lambda_{\sigma, \pi} = \int_{[0,1]^{p+q}} f(A_k, \tilde{A}) g(B) \cdot f(A_k, \tilde{C}) g(D) dt_1 \dots dt_{p+q} .$$

Note that $A_k \cup \tilde{A} \cup B = A_k \cup \tilde{C} \cup D = \{t_1, \dots, t_{p+q}\}$, $\tilde{A} \cap \tilde{C} = \emptyset$. Hence we have $\tilde{A} \cup B = \tilde{C} \cup D$. Since $\tilde{A} \cap \tilde{C} = \emptyset$, we have $\tilde{C} \subset B$ and $\tilde{A} \subset D$. From the fact that $(\tilde{A}, B)$ and $(\tilde{C}, D)$ are the partitions of the same set, we have $D \backslash \tilde{A} = B \backslash \tilde{C}$.

Hence we can write, with the obvious notations:

$$\lambda_{\sigma,\pi} =$$

$$= \int_{[0,1]^{p+q}} f(A_k, \tilde{A})g(\tilde{C}, B\backslash\tilde{C}) \cdot f(A_k, \tilde{C})g(\tilde{A}, D\backslash\tilde{A})dt_1 \ldots dt_{p+q}$$

$$= \int_{[0,1]^{p+q}} f(A_k, \tilde{A})g(\tilde{C}, B\backslash\tilde{C})f(A_k, \tilde{C})g(\tilde{A}, B\backslash\tilde{C})dA_k d\tilde{A}d\tilde{C}d(B\backslash\tilde{C})$$

$$= \int_{[0,1]^{q-p+2k}} (f \otimes_{p-k} g)(A_k, B\backslash\tilde{C})(f \otimes_{p-k} g)(A_k, B\backslash\tilde{C}) \cdot dA_k d(B\backslash\tilde{C})$$

$$= \|f \otimes_{p-k} g\|^2_{L^2([0,1]^{q-p+2k})}$$

where we have used the relation $D\backslash\tilde{A} = B\backslash\tilde{C}$ in the second line of the above equalities. Note that for $k = p$ we have $\lambda_{\sigma,\pi} = \|f \otimes g\|^2_{L^2}$. Hence we have

$$E[I_p^2(f)I_q^2(g)]$$

$$= p!\|f\|^2 \cdot q!\|g\|^2$$

$$\geq (p+q)! \left[ \frac{1}{((p+q)!)^2} \left[ \sum_{\sigma,\pi} \lambda_{\sigma,\pi}(k = p) + \sum_{\sigma,\pi} \lambda_{\sigma,\pi}(k \neq p)) \right] \right].$$

The number of $\lambda_{\sigma,\pi}$ with $(k = p)$ is exactly $\binom{p+q}{p}(p!)^2(q!)^2$, hence we have

$$p!q!\|f\|^2\|g\|^2 \geq p!q!\|f \otimes g\|^2 + \sum_{k=0}^{p-1} c_k\|f \otimes_{p-k} g\|^2_{L^2([0,1]^{q-p+2k})}$$

with $c_k > 0$. For this relation to hold we should have

$$\|f \otimes_{p-k} g\| = 0, \qquad k = 0, \ldots, p-1$$

in particular for $k = p - 1$, we have

$$\|f \otimes_1 g\| = 0.$$

($\Leftarrow$): From the Proposition 1, we see that it is sufficient to prove

$$((I + \mathcal{L})^{-1}e^{i\alpha F}\nabla F, \nabla I_q(g)) = 0 \text{ a.s.}$$

with $F = I_p(f)$, under the hypothesis $f \otimes_1 g = 0$ a.s.: Let us write $e^{i\alpha I_p(f)} = \sum_k I_k(h_k)$, then

$$e^{i\alpha I_p(f)}\nabla I_p(f) = p \cdot \sum_k I_k(h_k) \cdot I_{p-1}(f)$$

$$= p \cdot \sum_k \sum_{r=0}^{k\wedge(p-1)} \alpha_{p,k,r} I_{p-1+k-2r}(h_k \otimes_r f).$$

Hence

$$(I + \mathcal{L})^{-1}e^{i\alpha F}\nabla F = p \cdot \sum_k \sum_{r=0}^{k \wedge (p-1)} (1 + p + k - 1 - 2r)^{-1} I_{p-1+k-2r}(h_k \otimes_p f).$$

When we take the scalar product with $\nabla I_q(g)$, we will have terms of the type:

$$(I_{p-1+k-2r}(h_k \otimes_r f), I_{q-1}(g))_H =$$
$$= \sum_{i=1}^{\infty} I_{p-1+k-2r}(h_k \otimes_r f(e_i)) I_{q-1}(g(e_i)).$$

If we use the multiplication formula to calculate each term, we find the terms as

$$\sum_{i=1}^{\infty} \int (h_k \otimes_r f(e_i))(t_1, \ldots, t_{p+k-2r-1}) g(e_i)(t_1, \ldots, t_{q-1}) dt_1 dt_2 \ldots$$

$$= \int_{\theta=0}^{1} \int (h_k \otimes_r f(\theta))(t_1, \ldots, t_{p+k-2r-1}) g(\theta, t_1, \ldots, t_{q-1}) d\theta \, dt_1 \ldots$$

From the hypothesis we have

$$\int_0^1 f(\theta, t_1 \ldots) g(\theta, s_1 \ldots,) d\theta = 0 \quad \text{a.s.},$$

hence the Fubini theorem completes the proof. QED

**Corollary 1:** Let $f$ and $g$ be symmetric $L^2$-kernels respectively on $[0,1]^p$ and $[0,1]^q$. Let

$$S_f = \text{span}\{f \otimes_{p-1} h : h \in L^2([0,1])^{p-1}\}$$

and

$$S_g = \text{span}\{g \otimes_{q-1} k; k \in L^2(]0,1]^{q-1})\}.$$

Then the following are equivalent:

i) $I_p(f)$ and $I_q(g)$ are independent

ii) $S_f \perp S_g$ in $H$

iii) The Gaussian-generated $\sigma$-fields $\sigma\{I_1(k); k \in S_f\}$ and $\sigma\{I_1(l); l \in S_g\}$ are independent.

**Proof:** (i⇒ii): (i) implies that $f \otimes_1 g = 0$ a.s. If $a \in S_f$, $b \in S_g$ then $a = f \otimes_{p-1} h$ and $b = g \otimes_{q-1} k$ (rather finite sums of these kind of vectors). Then

$$(a,b) = (f \otimes_{p-1} h, g \otimes_{q-1} k) \quad = \quad (f \otimes_1 g, h \otimes k)_{(L^2)^{\otimes p+q-2}} \text{ (Fubini)}$$
$$= \quad 0.$$

(ii⇒i) If $(f \otimes_1 g, h \otimes k) = 0$ $\forall h \in L^2([0,1]^{p-1})$, $k \in L^2([0,1]^{q-1})$, then $f \otimes_1 g = 0$ a.s. since finite combinations of $h \otimes k$ are dense in $L^2([0,1]^{p+q-2})$.

(ii⇔iii)  Is obvious.                                                                    QED

**Proposition:**  Suppose that $I_p(f)$ is independent of $I_q(g)$ and $I_p(f)$ is independent of $I_r(h)$. Then $I_p(f)$ is independent of $\{I_q(g), I_r(h)\}$.

**Proof:**  We have $f \otimes_1 g = f \otimes_1 h = 0$ a.s.  This implies the independence of $I_p(f)$ and $\{I_q(g), I_r(h)\}$ from the calculations similar to those of the proof of sufficiency of the theorem.                                                         QED

In a similar way we have

**Proposition:**  Let $\{I_{p_\alpha}(f_\alpha); \alpha \in J\}$ and $I_{q_\beta}(g_\beta); \beta \in K\}$ be two arbitrary families of multiple Wiener integrals. The two families are independent if and only if $I_{p_\alpha}(f_\alpha)$ is independent of $I_{q_\beta}(g_\beta)$ for all $(\alpha, \beta) \in J \times K$.

**Corollary:**  If $I_p(f)$ and $I_q(g)$ are independent, so are also $I_p(f)(w + \tilde{h})$ and $I_q(g)(w + \tilde{k})$ for any $\tilde{h}, \tilde{k} \in H$.

**Proof:**  Let us denote, respectively, by $h$ and $k$ the Lebesgue densities of $\tilde{h}$ and $\tilde{k}$. We have then

$$I_p(f)(w + \tilde{h}) = \sum_{i=0}^{p} \binom{p}{i} (I_{p-i}(f), h^{\otimes i})_{H^{\otimes i}}.$$

Let us define $f[h^{\otimes i}] \in L^2[0,1]^{p-i}$ by

$$I_{p-i}(f[h^{\otimes i}]) = (I_{p-i}(f), h^{\otimes i}).$$

If $f \otimes_1 g = 0$ then it is easy to see that

$$f[h^{\otimes i}] \otimes_1 g[k^{\otimes j}] = 0,$$

hence the corollary follows from Theorem 1.

                                                                                       QED

From the corollary it follows

**Corollary:** $I_p(f)$ and $I_q(g)$ are independent if and only if the germ $\sigma$-fields

$$\sigma\{I_p(f), \nabla I_p(f), \ldots, \nabla^{p-1} I_p(f)\}$$

and

$$\sigma\{I_q(g), \ldots, \nabla^{q-1} I_q(g)\}$$

are independent.

**Corollary:** Let $X, Y \in L^2(\mu)$, $Y = \sum_0^\infty I_n(g_n)$. If

$$\nabla X \otimes_1 g_n = 0 \quad \text{a.s. } \forall n,$$

then $X$ and $Y$ are independent.

**Proof:** This follows from Prop. 1 $\hfill$ QED

**Corollary:** In particular, if $\tilde{h} \in H$, then $\nabla_{\tilde{h}} \varphi = 0$ a.s. implies that $\varphi$ and $I_1(h) = \delta \tilde{h}$ are independent.

# Chapter VIII

# Moment inequalities for Wiener functionals

In several applications, as limit theorems, large deviations, degree theory of Wiener maps, calculation of the Radon-Nikodym densities, etc., it is important to control the (exponential) moments of Wiener functionals by those of their derivatives. In this chapter we will give two results on this subject. The first one concerns the tail probabilities of the Wiener functionals with essentially bounded Gross-Sobolev derivatives. This result is a straightforward generalization of the celebrated Fernique's lemma which says that the square of the supremum of the Brownian path on any bounded interval has an exponential moment provided that it is multiplied with a sufficiently small, positive constant. The second inequality says that for a Wiener functional $F \in D_{p,1}$, we have

$$E_w \times E_z[U(F(w) - F(z))] \leq E_w \times E_z[U(\frac{\pi}{2} I_1(\nabla F(w))(z)],$$

where $w$ and $z$ represent two independent Wiener paths, $E_w$ and $E_z$ are the corresponding expectations, and $I_1(\nabla F(w))(z)$ is the first order Wiener integral with respect to $z$ of $\nabla F(w)$ and $U$ is any lower bounded, convex function on $\mathbf{R}$. Then combining these two inequalities we will obtain some interesting majorations.

The last inequality is an interpolation inequality which says that the Sobolev norm of first order can be majorated by the multiplication of the Sobolev norm of second order and of order zero.

# 1  Exponential tightness

First we will show the following result which is a consequence of the Doob inequality:

**Theorem 1:**   Let $\varphi \in D_{p,1}$ for some $p > 1$. Suppose that $\nabla\varphi \in L^{\infty}(\mu, H)$. Then we have

$$\mu\{|\varphi| > c\} \leq 2\exp-\frac{(c - E[\varphi])^2}{2\|\nabla\varphi\|^2_{L^{\infty}(\mu,H)}}\qquad\text{for any } c \geq 0 \,.$$

**Proof:**   Suppose that $E[\varphi] = 0$. Let $(e_i) \subset H$ be a complete, orthonormal basis of $H$. Define $V_n = \sigma\{\delta e_1, \ldots, \delta e_n\}$ and let $\varphi_n = E[P_{1/n}\varphi|V_n]$, where $P_t$ denotes the Ornstein- Uhlenbeck semigroup on $W$. Then, from Doob's Lemma,

$$\varphi_n = f_n(\delta e_1, \ldots, \delta e_n).$$

Note that, since $f_n \in \bigcap_{p,k} W_{p,k}(\mathbf{R}^n, \mu_n) \Rightarrow f_n$ is $C^{\infty}$ on $\mathbf{R}^n$ from the Sobolev injection theorem. Let $(B_t; t \in [0,1])$ be an $\mathbf{R}^n$-valued Brownian motion. Then

$$
\begin{aligned}
\mu\{|\varphi_n| > c\} &= \mathbf{P}\{|f_n(B_1)| > c\}\\
&\leq \mathbf{P}\{\sup_{t\in[0,1]} |E[f_n(B_1)|\mathcal{B}_t]| > c\}\\
&= \mathbf{P}\{\sup_{t\in[0,1]} |Q_{1-t}f_n(B_t)| > c\}\,,
\end{aligned}
$$

where $\mathbf{P}$ is the canonical Wiener measure on $C([0,1], \mathbf{R}^n)$ and $Q_t$ is the heat kernel associated to $(B_t)$, i.e.

$$Q_t(x, A) = \mathbf{P}\{B_t + x \in A\}\,.$$

From the Ito formula, we have

$$Q_{1-t}f_n(B_t) = Q_1 f_n(B_0) + \int_0^t (DQ_{1-s}f_n(B_s), dB_s)\,.$$

By definition

$$
\begin{aligned}
Q_1 f_n(B_0) &= Q_1 f_n(0) = \int f_n(y) \cdot Q_1(0, dy) =\\
&= \int f_n(y)e^{-\frac{|y|^2}{2}}\frac{dy}{(2\pi)^{n/2}}\\
&= E[E[P_{1/n}\varphi|V_n]]\\
&= E[P_{1/n}\varphi]\\
&= E[\varphi]\\
&= 0\,.
\end{aligned}
$$

Moreover we have $DQ_t f = Q_t Df$, hence

$$Q_{1-t}f_n(B_t) = \int_0^t (Q_{1-s}Df_n(B_s), dB_s) = M_t^n\,.$$

The Doob-Meyer process $\langle M^n, M^n \rangle_t$ of the martingale $M^n$ can be controlled as

$$\langle M^n, M^n \rangle_t = \int_0^t |DQ_{1-s} f_n(B_s)|^2 ds \leq$$

$$\leq \int_0^t \|Df_n\|^2_{C_b} ds = t\|\nabla f_n\|^2_{C_b} = t\|\nabla f_n\|_{L^\infty(\mu_n)} \leq$$

$$\leq t\|\nabla \varphi\|^2_{L^\infty(\mu, H)} .$$

Hence from the exponential Doob inequality, we obtain

$$P\{\sup_{t \in [0,1]} |Q_{1-t} f_n(B_t)| > c\} \leq 2\exp -\frac{c^2}{2\|\nabla \varphi\|^2_\infty} .$$

Hence

$$\mu\{|\varphi_n| > c\} \leq 2\exp \frac{-c^2}{2\|\nabla \varphi\|^2_{L^\infty(\mu, H)}} .$$

Since $\varphi_n \to \varphi$ in probability the proof is completed.     QED

**Corollary 1** Under the hypothesis of the theorem, for any $\lambda < \frac{1}{2\|\nabla \varphi\|_\infty}$, we have

$$E[\exp \lambda |\varphi|^2] < \infty.$$

**Proof:** The first part follows from the fact that, for $F > 0$ a.s.,

$$E[F] = \int_0^\infty P\{F > t\} dt .$$

The second part follows from the fact that $|\nabla \|w\|\|_H \leq 1$.     QED

**Remark:** In the next section we will give a precise estimation for $E[\exp \lambda F^2]$.

In the applications, we encounter random variables $F$ satisfying

$$|F(w+h) - F(w)| \leq c|h|_H,$$

almost surely, for any $h$ in the Cameron-Martin space $H$ and a fixed constant $c > 0$, without any hypothesis of integrability. For example, $F(w) = \sup_{t \in [0,1]} |w(t)|$, defined on $C_0[0,1]$ is such a functional. In fact the above hypothesis contains the integrability and Sobolev differentiability of $F$. We begin first by proving that under the integrability hypothesis, such a functional is in the domain of $\nabla$:

**Lemma 1** Suppose that $F : W \mapsto \mathbf{R}$ is a measurable random variable in $\cup_{p>1} L^p(\mu)$, satisfying

$$|F(w + h) - F(w)| \le c|h|_H,$$

almost surely, for any $h \in H$, where $c > 0$ is a fixed constant. Then $F$ belongs to $D_{p,1}$ for any $p > 1$.

**Remark:** We call such a functional H-Lipschitz.

**Proof:** Since, for some $p_0 > 1$, $F \in L^{p_0}$, the distributional derivative of $F$, $\nabla F$ exists . We have $\nabla_k F \in D'$ for any $k \in H$. Moreover, for $\phi \in D$, from the integration by parts formula

$$
\begin{aligned}
E[\nabla_k F \, \phi] &= -E[F \, \nabla_k \phi] + E[F \delta k \, \phi] \\
&= -\frac{d}{dt}|_{t=0} E[F \, \phi(w + tk)] + E[F \delta k \, \phi] \\
&= -\frac{d}{dt}|_{t=0} E[F(w - tk) \, \phi \, \varepsilon(t\delta k)] + E[F \delta k \, \phi] \\
&= \lim_{t \to 0} -E[\frac{F(w - tk) - F(w)}{t} \phi],
\end{aligned}
$$

where $\varepsilon(\delta k) = \exp \delta k - 1/2|k|^2$. Consequently,

$$
\begin{aligned}
|E[\nabla_k F \, \phi]| &\le c|k|_H E[|\phi|] \\
&\le c|k|_H \|\phi\|_q,
\end{aligned}
$$

for any $q > 1$, i.e., $\nabla F$ belongs to $L^p(\mu, H)$ for any $p > 1$. Let now $(e_i; i \in \mathbf{N})$ be a complete, orthonormal basis of $H$, denote by $V_n$ the sigma-field generated by $\delta e_1, \dots, \delta e_n$, $n \in \mathbf{N}$ and let $\pi_n$ be the orthogonal projection onto the the subspace of $H$ spanned by $e_1, \dots, e_n$. Let us define

$$F_n = E[P_{1/n} F | V_n],$$

where $P_{1/n}$ is the Ornstein-Uhlenbeck semigroup at the instant $t = 1/n$. Then $F_n \in \cap_k D_{p_0, k}$ and it is immediate, from the martingale convergence theorem and from the fact that $\pi_n$ tends to the identity operator of $H$ in the norm-topology, that

$$\nabla F_n = E[e^{-1/n} \pi_n P_{1/n} \nabla F | V_n] \to \nabla F,$$

in $L^p(\mu, H)$, for any $p > 1$, as $n$ tends to infinity. Since, by construction, $(F_n; n \in \mathbf{N})$ converges also to $F$ in $L^{p_0}(\mu)$, $F$ belongs to $D_{p_0, 1}$. Hence we can apply the Corollary 1.

                                                                          Q.E.D.

**Lemma 2** Suppose that $F : W \mapsto \mathbf{R}$ is a measurable random variable satisfying

$$|F(w + h) - F(w)| \le c|h|_H,$$

almost surely, for any $h \in H$, where $c > 0$ is a fixed constant. Then $F$ belongs to $D_{p,1}$ for any $p > 1$.

**Proof:** Let $F_n = |F| \wedge n$, $n \in \mathbf{N}$. A simple calculation shows that

$$|F_n(w+h) - F_n(w)| \leq c|h|_H,$$

hence $F_n \in D_{p,1}$ for any $p > 1$ and $|\nabla F_n| \leq c$ almost surely from the Lemma 1. We have from the Ito-Clark formula (cf. Chapter V),

$$F_n = E[F_n] + \int_0^1 E[D_s F_n | \mathcal{F}_s] dW_s.$$

From the definition of the stochastic integral, we have

$$
\begin{aligned}
E\left[(\int_0^1 E[D_s F_n | \mathcal{F}_s] dW_s)^2\right] &= E\left[\int_0^1 |E[D_s F_n | \mathcal{F}_s]|^2 ds\right] \\
&\leq E\left[\int_0^1 |D_s F_n|^2 ds\right] \\
&= E[|\nabla F_n|^2] \\
&\leq c^2
\end{aligned}
$$

Since $F_n$ converges to $|F|$ in probability, and the stochastic integral is bounded in $L^2(\mu)$, by taking the difference, we see that $(E[F_n], n \in \mathbf{N})$ is a sequence of (degenerate) random variables bounded in the space of random variables under the topology of convergence in probability, denoted by $L^0(\mu)$. Therefore $\sup_n \mu\{E[F_n] > c\} \to 0$ as $c \to \infty$. Hence $\lim_n E[F_n] = E[|F|]$ is finite. Now we apply the dominated convergence theorem to obtain that $F \in L^2(\mu)$. Since the distributional derivative of $F$ is a square integrable random variable, $F \in D_{2,1}$. We can now apply the Lemma 1 which implies that $F \in D_{p,1}$ for any $p$.

$$\text{Q.E.D.}$$

**Remark:** Although we have used the classical Wiener space structure in the proof, the case of the Abstract Wiener space can be reduced to this case using the method explained in the appendix of Chapter IV.

**Corollary (Fernique's Lemma): 2** For any $\lambda < \frac{1}{2}$, we have

$$E[\exp \lambda \|w\|_W^2] < \infty,$$

where $\|w\|$ is the norm of the Wiener path $w \in W$.

We will see another application of this result later.

# 2 Coupling inequalities

We begin with the following elementary lemma (cf. [18]):

**Lemma:3** Let $X$ be a Gaussian r.v. on $\mathbf{R}^d$. Then for any convex function $U$ on $\mathbf{R}$ and $C^1$-function $V : \mathbf{R}^d \to \mathbf{R}$, we have the following inequality:

$$E[U(V(X) - V(Y))] \leq E\left[U\left(\frac{\pi}{2}(V'(X), Y)_{\mathbf{R}^d}\right)\right],$$

where $Y$ is an independent copy of $X$ and $E$ is the expectation with respect to the product measure.

**Proof:** Let $X_\theta = X \sin\theta + Y \cos\theta$. Then

$$V(X) - V(Y) = \int_0^{\pi/2} \frac{d}{d\theta} V(X_\theta) d\theta$$

$$= \int_0^{\pi/2} (V'(X_\theta), X_\theta')_{\mathbf{R}^d} \, d\theta$$

$$= \frac{\pi}{2} \int_0^{\pi/2} (V'(X_\theta), X_\theta')_{\mathbf{R}^d} \, d\tilde{\theta}$$

where $d\tilde{\theta} = \frac{d\theta}{\pi/2}$. Since $U$ is convex, we have

$$U(V(X) - V(Y)) \leq \int_0^{\pi/2} U\left(\frac{\pi}{2}(V'(X_\theta), X_\theta')\right) d\tilde{\theta}.$$

Moreover $X_\theta$ and $X_\theta'$ are two independent Gaussian random variables with the same law as the one of $X$. Hence

$$E[U(V(X) - V(Y))] \leq \int_0^{\pi/2} E\left[U\left(\frac{\pi}{2}(V'(X), Y)\right)\right] d\tilde{\theta}$$

$$= E\left[U\left(\frac{\pi}{2}(V'(X), Y)\right)\right]. \qquad \text{QED}$$

Now we will extend this result to the Wiener space:

**Theorem 2** Suppose that $\varphi \in D_{p,1}$, for some $p > 1$ and $U$ is a lower bounded, convex function (hence lower semi-continuous) on $\mathbf{R}$. We have

$$E[U(\varphi(w) - \varphi(z))] \leq E\left[U\left(\frac{\pi}{2} I_1(\nabla\varphi(w))(z)\right)\right]$$

where $E$ is taken with respect to $\mu(dw) \times \mu(dz)$ on $W \times W$ and on the classical Wiener space, we have

$$I_1(\nabla\varphi(w))(z) = \int_0^1 \frac{d}{dt} \nabla\varphi(w, t) dz_t.$$

**Proof:** Suppose first that

$$\varphi = f(\delta h_1(w), \ldots, \delta h_n(w))$$

with $f$ smooth on $\mathbf{R}^n$, $h_i \in H$, $(h_i, h_j) = \delta_{ij}$. We have

$$
\begin{aligned}
I_1(\nabla\varphi(w))(z) &= I_1\Big(\sum_{i=1}^n \partial_i f(\delta h_1(w), \ldots, \delta h_n(w))h_i\Big) \\
&= \sum_{i=1}^n \partial_i f(\delta h_1(w), \ldots, \delta h_n(w))I_1(h_i)(z) \\
&= (f'(X), Y)_{\mathbf{R}^n}
\end{aligned}
$$

where $X = (\delta h_1(w), \ldots, \delta h_n(w))$ and $Y = (\delta h_1(z), \ldots, \delta h_n(z))$. Hence the inequality is trivially true in this case.

For general $\varphi$, let $(h_i)$ be a complete, orthonormal basis in $H$,

$$V_n = \sigma\{\delta h_1, \ldots, \delta h_n\}$$

and let

$$\varphi_n = E[P_{1/n}\varphi|V_n],$$

where $P_{1/n}$ is the Ornstein-Uhlenbeck semigroup on $W$.

We have then

$$E[U(\varphi_n(w) - \varphi_n(z))] \le E\Big[U\Big(\frac{\pi}{2}I_1(\nabla\varphi_n(w))(z)\Big)\Big].$$

Let $\pi_n$ be the orthogonal projection from $H$ onto span $\{h_1, \ldots, h_n\}$. We have

$$
\begin{aligned}
I_1(\nabla\varphi_n(w))(z) &= I_1(\nabla_w E_w[P_{1/n}\varphi|V_n])(z) \\
&= I_1(E_w[e^{-1/n}P_{1/n}\pi_n\nabla\varphi|V_n])(z) \\
&= I_1(\pi_n E_w[e^{-1/n}P_{1/n}\nabla\varphi|V_n])(z) \\
&= E_z[I_1^z(E_w[e^{-1/n}P_{1/n}^w\nabla\varphi|V_n])|\tilde{V}_n]
\end{aligned}
$$

where $\tilde{V}_n$ is the copy of $V_n$ on the second Wiener space. Then

$$
\begin{aligned}
E\Big[&U\Big(\frac{\pi}{2}I_1(\nabla\varphi_n(w))(z)\Big)\Big] \\
&\le E\Big[U\Big(\frac{\pi}{2}I_1(E_w[e^{-1/n}P_{1/n}\nabla\varphi|V_n])(z)\Big)\Big] \\
&= E\Big[U\Big(\frac{\pi}{2}e^{-1/n}E_w[I_1(P_{1/n}\nabla\varphi(w))(z)|V_n]\Big)\Big] \\
&\le E\Big[U\Big(\frac{\pi}{2}e^{-1/n}I_1(P_{1/n}\nabla\varphi(w))(z)\Big)\Big] \\
&= E\Big[U\Big(\frac{\pi}{2}e^{-1/n}P_{1/n}^w I_1(\nabla\varphi(w))(z)\Big)\Big] \\
&\le E\Big[U\Big(\frac{\pi}{2}e^{-1/n}I_1(\nabla\varphi(w))(z)\Big)\Big] \\
&= E\Big[U\Big(\frac{\pi}{2}P_{1/n}^{(z)}I_1(\nabla\varphi(w))(z)\Big)\Big] \\
&\le E\Big[U\Big(\frac{\pi}{2}I_1(\nabla\varphi(w))(z)\Big)\Big].
\end{aligned}
$$

Now Fatou's lemma completes the proof.                                    QED

Let us give some consequences of this result:

**Theorem 3**   The following Poincaré inequalities are valid:

i) $E[\exp(\varphi - E[\varphi])] \leq E\left[\exp\frac{\pi^2}{8}|\nabla\varphi|^2_H\right]$,

ii) $E[|\varphi - E[\varphi]|] \leq \frac{\pi}{2}E[|\nabla\varphi|_H]$.

iii) $E[|\varphi - E[\varphi]|^{2k}] \leq \left(\frac{\pi}{2}\right)^{2k}\frac{(2k)!}{2^k k!}E[|\nabla\varphi|^{2k}_H]$, $k \in \mathbf{N}$.

**Remark:**   Let us note that the result of (ii) can not be obtained with the classical methods, such as the Ito-Clark representation theorem, since the optional projection is not a continuous map in $L^1$-setting. Moreover, using the Hölder inequality and the Stirling formula, we deduce the following set of inequalities:

$$\|\varphi - E[\varphi]\|_p \leq p\frac{\pi}{2}\|\nabla\varphi\|_{L^p(\mu,H)},$$

for any $p \geq 1$ . To compare this result with those already known, let us recall that using first the Ito-Clark formula, then the Burkholder-Davis-Gundy inequality combined with the convexity inequalities for the dual projections and some duality techniques, we obtain, only for $p > 1$ the inequality

$$\|\varphi - E[\varphi]\|_p \leq Kp^{3/2}\|\nabla\varphi\|_{L^p(\mu,H)},$$

where $K$ is some positive constant.

**Proof:**   Replacing the function $U$ of Theorem 2 by the exponential function, we have

$$
\begin{aligned}
E[\exp(\varphi - E[\varphi])] &\leq E_w \times E_z[\exp(\varphi(w) - \varphi(z))] \leq \\
&\leq E_w\left[E_z\left[\left[\exp\frac{\pi}{2}I_1(\nabla\varphi(w))(z)\right]\right]\right] \\
&= E\left[\exp\frac{\pi}{2}|\nabla\varphi(w)|^2_H\right].
\end{aligned}
$$

(ii) and (iii) are similar with $U(x) = |x|^k$. $k \in \mathbf{N}$.                QED

**Theorem 4:**   Let $\varphi \in D_{p,2}$ for some $p > 1$ and that $\nabla|\nabla\varphi|_H \in L^\infty(\mu, H)$ (in particular, this is satisfied if $\nabla^2\varphi \in L^\infty(\mu, H\tilde{\otimes}_2 H)$). Then there exists some $\lambda > 0$ such that

$$E[\exp\lambda|\varphi|] < \infty.$$

**Proof:** From Theorem 3, (i), we know that

$$E[\exp \lambda |\varphi - E[\varphi]|] \leq 2E\left[\exp \frac{\lambda^2 \pi^2}{8}|\nabla \varphi|^2\right]$$

. Hence it is sufficient to prove that

$$E[\exp \lambda^2 |\nabla \varphi|^2] < \infty$$

for some $\lambda > 0$. However Theorem 1 applies since $\nabla |\nabla \varphi| \in L^{\infty}(\mu, H)$.

QED

**Corollary 3** Let $F \in D_{p,1}$ for some $p > 1$ such that $|\nabla F|_H \in L^{\infty}(\mu)$. We then have

$$E[\exp \lambda F^2] \leq E\left[\frac{1}{\sqrt{1 - \frac{\lambda \pi^2}{4}|\nabla F|_H^2}} \exp\left(\frac{\lambda E[F]^2}{1 - \frac{\lambda \pi^2}{4}|\nabla F|^2}\right)\right],$$

for any $\lambda > 0$ such that $\||\nabla F|_H\|_{L^{\infty}(\mu)}^2 \frac{\lambda \pi^2}{4} < 1$.

**Proof:** Ley $Y$ be an auxiliary, real-valued Gaussian random variable, living on a separate probability space $(\Omega, \mathcal{U}, P)$ with variance one and zero expectation. We have, using Corollary 2 :

$$
\begin{aligned}
E[\exp \lambda F^2] &= E \otimes E_P[\exp \sqrt{2\lambda}FY] \\
&\leq E \otimes E_P\left[\exp\{\sqrt{2\lambda}E[F]Y + |\nabla F|^2 Y^2 \frac{\lambda \pi^2}{4}\}\right] \\
&= E\left[\frac{1}{\sqrt{1 - \frac{\lambda \pi^2}{2}|\nabla F|_H^2}} \exp\left(\frac{\lambda E[F]^2}{1 - \frac{\lambda \pi^2}{2}|\nabla F|^2}\right)\right],
\end{aligned}
$$

where $E_P$ denotes the expectation with respect to the probability $P$.

Q.E.D.

# 3    An interpolation inequality

Another useful inequality for the Wiener functionals * is the following interpolation inequality which helps to control the $L^p$- norm of $\nabla F$ with the help of the $L^p$-norms of $F$ and $\nabla^2 F$.

---

*This result has been proven as an answer to a question posed by D. W. Stroock, cf. also [4]

**Theorem 5:**   For any $p > 1$, there exists a constant $C_p$, such that, for any $F \in D_{p,2}$, one has

$$\|\nabla F\|_p \le C_p \left[\|F\|_p + \|F\|_p^{1/2}\|\nabla^2 F\|_p^{1/2}\right].$$

The theorem, will be proven, thanks to the Meyer inequalities, if we can prove the following

**Theorem 6:**   For any $p > 1$, we have

$$\|(I + \mathcal{L})^{1/2} F\|_p \le \frac{4}{\Gamma(1/2)} \|F\|_p^{1/2}\|(I + \mathcal{L})F\|_p^{1/2}.$$

**Proof:**   Denote by $G$ the functional $(I + \mathcal{L})F$. Then we have $F = (I + \mathcal{L})^{-1}G$. Therefore it suffices to show that

$$\|(I + \mathcal{L})^{-1/2} G\|_p \le \frac{4}{\Gamma(1/2)} \|G\|_p^{1/2}\|(I + \mathcal{L})^{-1}G\|_p^{1/2}.$$

We have

$$(I + \mathcal{L})^{-1/2}G = \frac{\sqrt{2}}{\Gamma(1/2)} \int_0^\infty t^{-1/2} e^{-t} P_t G \, dt,$$

where $P_t$ denotes the semigroup of Ornstein-Uhlenbeck. For any $a > 0$, we can write

$$(I + \mathcal{L})^{-1/2}G = \frac{\sqrt{2}}{\Gamma(1/2)} \left[\int_0^a t^{-1/2} e^{-t} P_t G \, dt + \int_a^\infty t^{-1/2} e^{-t} P_t G \, dt\right].$$

Let us denote the two terms at the right hand side of the above equality, respectively, by $I_a$ and $II_a$. We have

$$\|(I + \mathcal{L})^{-1/2}G\|_p \le \frac{\sqrt{2}}{\Gamma(1/2)} [\|I_a\|_p + \|II_a\|_p].$$

The first term at the right hand side can be majorated as

$$\begin{aligned}
\|I_a\|_p &\le \int_0^a t^{-1/2}\|G\|_p \, dt \\
&= 2\sqrt{a}\|G\|_p.
\end{aligned}$$

Let $g = (I + \mathcal{L})^{-1}G$. Then

$$\begin{aligned}
\int_a^\infty t^{-1/2} e^{-t} P_t G \, dt &= \int_a^\infty t^{-1/2} e^{-t} P_t (I + \mathcal{L})(I + \mathcal{L})^{-1}G \, dt \\
&= \int_a^\infty t^{-1/2} e^{-t} P_t (I + \mathcal{L}) g \, dt \\
&= \int_a^\infty t^{-1/2} \frac{d}{dt}(e^{-t} P_t) \, dt \\
&= -a^{-1/2} e^{-a} P_a g + \frac{1}{2}\int_a^\infty t^{-3/2} e^{-t} P_t g \, dt,
\end{aligned}$$

where the third equality follows from the integration by parts formula. Therefore

$$
\begin{aligned}
\|II_a\|_p &\leq a^{-1/2}\|e^{-a}P_a g\|_p + \frac{1}{2}\int_a^\infty t^{-3/2}\|e^{-t}P_t g\|_p\, dt \\
&\leq a^{-1/2}\|g\|_p + \frac{1}{2}\int_a^\infty t^{-3/2}\|g\|_p\, dt \\
&= 2a^{-1/2}\|g\|_p \\
&= 2a^{-1/2}\|(I+\mathcal{L})^{-1}G\|_p.
\end{aligned}
$$

Finally we have

$$
\|(I+\mathcal{L})^{-1/2}G\|_p \leq \frac{2}{\Gamma(1/2)}[a^{1/2}\|G\|_p + a^{-1/2}\|(I+\mathcal{L})^{-1}G\|_p].
$$

This expression attains its minimum when we take

$$
a = \frac{\|(I+\mathcal{L})^{-1}G\|_p}{\|G\|_p}.
$$

Q.E.D.

Combining this theorem with Meyer inequalities, we have

**Corollary 4** Suppose that $(F_n, n \in \mathbf{N})$ converges to zero in $D_{p,k}$, $p > 1, k \in \mathbf{Z}$, and that it is bounded in $D_{p,k+2}$. Then the convergence takes place also in $D_{p,k+1}$.

# Chapter IX

# Introduction to the theorem of Ramer

The Girsanov theorem tells us that if $u : W \mapsto H$ is a Wiener functional such that $\frac{du}{dt} = \dot{u}(t)$ is an adapted process such that

$$E[\exp - \int_0^1 \dot{u}(s)dW_s - \frac{1}{2}\int_0^1 |\dot{u}(s)|^2 ds] = 1,$$

then under the new probability $Ld\mu$, where

$$L = \exp - \int_0^1 \dot{u}(s)dW_s - \frac{1}{2}\int_0^1 |\dot{u}(s)|^2 ds,$$

$w+u(w)$ is a Brownian motion. The theorem of Ramer studies the same problem without hypothesis of adaptedness of the process $\dot{u}$. This problem has been initiated by Cameron and Martin. Their work has extended by Gross and others. It was Ramer [19] who gave a main impulse to the problem by realizing that the ordinary determinant can be replaced by the modified Carleman-Fredholm determinant via defining a Gaussian divergence instead of the ordinary Lebesgue divergence. The problem has been further studied by Kusuoka [11] and the final solution in the case of (locally) differentiable shifts in the Cameron-Martin space direction has been given by Üstünel and Zakai [31]. In this chapter we will give a partial ( however indispensable for the proof of the general ) result.

To understand the problem, let us consider first the finite dimensional case: let $W = \mathbf{R}^n$ and let $\mu_n$ be the standard Gauss measure on $\mathbf{R}^n$. If $u : \mathbf{R}^n \mapsto \mathbf{R}^n$ is a differentiable mapping such that $I + u$ is a diffeomorphism of $\mathbf{R}^n$, then the theorem of Jacobi tells us that, for any smooth function $F$ on $\mathbf{R}^n$, we have

$$\int_{\mathbf{R}^n} F(x + u(x))|\det(I + \partial u(x))|\exp\{- <u(x), x> -\frac{1}{2}|u|^2\}\mu_n(dx)$$

$$= \int_{\mathbf{R}^n} F(x)\mu_n(dx),$$

where $\partial u$ denotes the derivative of $u$. The natural idea now is to pass to the infinite dimension. For this, note that, if we define $\det_2(I + \partial u)$ by

$$\begin{aligned}
\det_2(I + \partial u(x)) &= \det(I + \partial u(x)) \exp{-\text{trace}\,\partial u(x)} \\
&= \prod_i (1 + \lambda_i) \exp{-\lambda_i},
\end{aligned}$$

where $(\lambda_i)$ are the eigenvalues of $\partial u(x)$ counted with respect to their multiplicity, then the density of the left hand side can be written as

$$\Lambda = |\det_2(I + \partial u(x))| \exp{- <u(x), x> +\text{trace}\,\partial u(x) - \frac{1}{2}|u|^2}$$

and let us remark that

$$<u(x), x> -\text{trace}\,\partial u(x) = \delta u(x),$$

where $\delta$ is the adjoint of the $\partial$ with respect to the Gaussian measure $\mu_n$. Hence, we can express the density $\Lambda$ as

$$\Lambda = |\det_2(I + \partial u(x))| \exp{-\delta u(x) - \frac{|u(x)|^2}{2}}.$$

As remarked first by Ramer, cf.[19], this expression has two advantages: first $\det_2(I + \partial u)$, called Carleman-Fredholm determinant, can be defined for the mappings $u$ such that $\partial u(x)$ is with values in the space of Hilbert-Schmidt operators rather than nuclear operators (the latter is a smaller class than the former), secondly, as we have already seen, $\delta u$ is well-defined for a large class of mappings meanwhile $<u(x), x>$ is a highly singular object in the Wiener space.

After all these remarks, we can announce the main result of this chapter, using our standard notations, as

**Theorem:**  Suppose that $u : W \mapsto H$ is a measurable map belonging to $D_{p,1}(H)$ for some $p > 1$. Assume that there are constants $c$ and $d$ with $c < 1$ such that for almost all $w \in W$,

$$\|\nabla u\| \leq c < 1$$

and

$$\|\nabla u\|_2 \leq d < \infty,$$

where $\| \cdot \|$ denotes the operator norm and $\| \cdot \|_2$ denotes the Hilbert-Schmidt norm for the linear operators on $H$. Then:

- Almost surely $w \mapsto T(w) = w + u(w)$ is bijective. The inverse of $T$, denoted by $S$ is of the form $S(w) = w + v(w)$, where $v$ belongs to $D_{p,1}(H)$ for any $p > 1$, moreover

$$\|\nabla v\| \leq \frac{c}{1 - c} \quad \text{and} \quad \|\nabla v\|_2 \leq \frac{d}{1 - c},$$

  $\mu$-almost surely.

- For all bounded and measurable $F$, we have

$$E[F(w)] = E[F(T(w)) \cdot |\Lambda_u(w)|]$$

and in particular

$$E|\Lambda_u| = 1,$$

where

$$\Lambda_u = |\det{}_2(I + \nabla u)| \exp -\delta u - \frac{1}{2}|u|_H^2,$$

and $\det_2(I + \nabla u)$ denotes the Carleman-Fredholm determinant of $I + \nabla u$.

- The measures $\mu$, $T^\star\mu$ and $S^\star\mu$ are mutually absolutely continuous, where $T^\star\mu$ (respectively $S^\star\mu$) denotes the image of $\mu$ under $T$ (respectively $S$). We have

$$\frac{dS^\star\mu}{d\mu} = |\Lambda_u|,$$
$$\frac{dT^\star\mu}{d\mu} = |\Lambda_v|,$$

where $\Lambda_v$ is defined similarly.

**Remark**   As it has been remarked in [15], if $\|\nabla u\| \leq 1$ instead of $\|\nabla u\| \leq c < 1$, then taking $u_\epsilon = (1 - \epsilon)u$ we see that the hypothesis of the theorem are satisfied for $u_\epsilon$. Hence using the Fatou lemma, we obtain

$$E[F \circ T \, |\Lambda_u|] \leq E[F]$$

for any positive $F \in C_b(W)$. Consequently, if $\Lambda_u \neq 0$ almost surely, then $T^\star\mu$ is absolutely continuous with respect to $\mu$.

The proof of the theorem will be done in several steps. As we have indicated above, the main idea is to pass to the limit from finite to infinite dimensions. The key point in this procedure will be the use of the Theorem 1 of the preceding chapter which will imply the uniform integrability of the finite dimensional densities. We shall first prove the same theorem in the cylindrical case:

**Lemma 1**   Let $\xi : W \longmapsto H$ be a shift of the following form:

$$\xi(w) = \sum_{i=1}^{n} \alpha_i(\delta h_1, \dots, \delta h_n)h_i,$$

with $\alpha_i \in C^\infty(\mathbf{R}^n)$ with bounded first derivative, $h_i \in W^*$ are orthonormal* in $H$. Suppose furthermore that $\|\nabla\xi\| \leq c < 1$ and that $\|\nabla\xi\|_2 \leq d$ as above. Then we have

---

*In fact $h_i \in W^*$ should be distinguished from its image in $H$, denoted by $j(h)$. For notational simplicity, we denote both by $h_i$, as long as there is no ambiguity.

- Almost surely $w \mapsto U(w) = w + \xi(w)$ is bijective.

- The measures $\mu$ and $U^*\mu$ are mutually absolutely continuous.

- For all bounded and measurable $F$, we have

$$E[F(w)] = E[F(U(w)) \cdot |\Lambda_\xi(w)|]$$

for all bounded and measurable $F$ and in particular

$$E[|\Lambda_\xi|] = 1,$$

where

$$\Lambda_\xi = |\det_2(I + \nabla\xi)| \exp -\delta\xi - \frac{1}{2}|\xi|_H^2.$$

- The inverse of $U$, denoted by $V$ is of the form $V(w) = w + \eta(w)$, where

$$\eta(w) = \sum_{i=1}^{n} \beta_i(\delta h_1, \ldots, \delta h_n)h_i,$$

such that $\|\nabla\eta\| \leq \frac{c}{1-c}$ and $\|\nabla\eta\|_2 \leq \frac{d}{1-c}$.

**Proof:** Note first that due to the Corollary 1 of the Chapter VIII, $E[\exp \lambda|\xi|^2] < \infty$ for any $\lambda < \frac{1}{2c}$. We shall construct the inverse of $U$ by imitating the fixed point techniques: let

$$\begin{aligned} \eta_0(w) &= 0 \\ \eta_{n+1}(w) &= -\xi(w + \eta_n(w)). \end{aligned}$$

We have

$$\begin{aligned} |\eta_{n+1}(w) - \eta_n(w)|_H &\leq c|\eta_n(w) - \eta_{n-1}(w)|_H \\ &\leq c^n|\xi(w)|_H. \end{aligned}$$

Therefore $\eta(w) = \lim_{n\to\infty} \eta_n(w)$ exists and it is majorated by $\frac{1}{1-c}|\xi(w)|$. By the triangle inequality

$$\begin{aligned} |\eta_{n+1}(w + h) - \eta_{n+1}(w)| &\leq |\xi(w + h + \eta_n(w + h)) - \xi(w + \eta_n(w))| \\ &\leq c|h| + c|\eta_n(w + h) - \eta_n(w)|. \end{aligned}$$

Hence passing to the limit, we find

$$|\eta(w + h) - \eta(w)| \leq \frac{c}{1 - c}|h|.$$

We also have

$$\begin{aligned} U(w + \eta(w)) &= w + \eta(w) + \xi(w + \eta(w)) \\ &= w + \eta(w) - \eta(w) \\ &= w, \end{aligned}$$

hence $U \circ (I_W + \eta) = I_W$, i.e., $U$ is an onto map. If $U(w) = U(w')$, then

$$
\begin{aligned}
|\xi(w) - \xi(w')| &= |\xi(w' + \xi(w') - \xi(w)) - \xi(w')| \\
&\leq c|\xi(w) - \xi(w')|,
\end{aligned}
$$

which implies that $U$ is also injective. To show the Girsanov identity, let us complete the sequence $(h_i; i \leq n)$ to a complete orthonormal basis whose elements are chosen from $W^*$. From a theorem of Ito-Nisio [8], we can express the Wiener path $w$ as

$$
w = \sum_{i=1}^{\infty} \delta h_i(w) h_i,
$$

where the sum converges almost surely in the norm topology of $W$. Let $F$ be a nice function on $W$, denote by $\mu_n$ the image of the Wiener measure $\mu$ under the map $w \mapsto \sum_{i \leq n} \delta h_i(w) h_i$ and by $\nu$ the image of $\mu$ under $w \mapsto \sum_{i > n} \delta h_i(w) h_i$. Evidently $\mu = \mu_n \times \nu$. Therefore

$$
\begin{aligned}
E[F \circ U \,|\Lambda_\xi|] &= \int_{\mathbf{R}^n} E_\nu[F'(w + \sum_{i \leq n}(x_i + \alpha_i(x_1 \ldots, x_n))h_i) |\Lambda_\xi|] \mu_{\mathbf{R}^n}(dx) \\
&= E[F],
\end{aligned}
$$

where $\mu_{\mathbf{R}^n}(dx)$ denotes the standard Gaussian measure on $\mathbf{R}^n$ and the equality follows from the Fubini theorem. In fact by changing the order of integrals, we reduce the problem to a finite dimensional one and then the result is immediate from the theorem of Jacobi as explained above. From the construction of $V$, it is trivial to see that

$$
\eta(w) = \sum_{i \leq n} \beta_i(\delta h_1, \ldots, \delta h_n) h_i,
$$

for some vector field $(\beta_1, \ldots, \beta_n)$ which is a $C^\infty$ mapping from $\mathbf{R}^n$ into itself due to the finite dimensional inverse mapping theorem. Now it is routine to verify that

$$
\nabla \eta = -(I + \nabla \eta)^* \nabla \xi \circ V,
$$

hence

$$
\begin{aligned}
\|\nabla \eta\|_2 &\leq \|I + \nabla \eta\| \|\nabla \xi \circ V\|_2 \\
&\leq (1 + \|\nabla \eta\|) \|\nabla \xi \circ V\|_2 \\
&\leq (1 + \frac{c}{1-c})d \\
&= \frac{d}{1-c}.
\end{aligned}
$$

Q.E.D.

**Lemma 2** With the notations and hypothesis of Lemma 1, we have

$$
\delta \xi \circ V = -\delta \eta - |\eta|_H^2 + \mathrm{trace}(\nabla \xi \circ V) \cdot \nabla \eta,
$$

almost surely.

**Proof:** We have

$$\delta\xi = \sum_{i=1}^{\infty}(\xi, e_i)\delta e_i - \nabla_{e_i}(\xi, e_i),$$

where the sum converges in $L^2$ and the result is independent of the choice of the orthonormal basis $(e_i;\ i \in \mathbf{N})$. Therefore we can choose as basis $h_1, \ldots, h_n$ that we have already used in Lemma 1, completed with the elements of $W^*$ to form an orthonormal basis of $H$, denoted by $(h_i;\ i \in \mathbf{N})$. Hence

$$\delta\xi = \sum_{i=1}^{n}(\xi, h_i)\delta h_i - \nabla_{h_i}(\xi, h_i).$$

From the Lemma 1, we have $\xi \circ V = -\eta$ and since, $h_i$ are originating from $W^*$, it is immediate to see that $\delta h_i \circ V = \delta h_i + (h_i, \eta)$. Moreover, from the preceding lemma we know that $\nabla(\xi \circ V) = (I + \nabla\eta)^* \nabla\xi \circ V$. Consequently, applying all this, we obtain

$$
\begin{aligned}
\delta\xi \circ V &= \sum_{1}^{n}(\xi \circ V, h_i)(\delta h_i + (h_i, \eta)) - (\nabla_{h_i}(\xi, h_i)) \circ V \\
&= (\xi \circ V, \eta) + \delta(\xi \circ V) + \sum_{1}^{n}\nabla_{h_i}(\xi \circ V, h_i) - \nabla_{h_i}(\xi, h_i) \circ V \\
&= -|\eta|^2 - \delta\eta + \sum_{1}^{n}(\nabla\xi \circ V [h_i], \nabla\eta [h_i]) \\
&= -|\eta|^2 - \delta\eta + \operatorname{trace}(\nabla\xi \circ V \cdot \nabla\eta),
\end{aligned}
$$

where $\nabla\xi [h]$ denotes the Hilbert-Schmidt operator $\nabla\xi$ applied to the vector $h \in H$.                                                                   Q.E.D.

**Remark**   Since $\xi$ and $\eta$ are symmetric, we have $\eta \circ U = -\xi$ and consequently

$$\delta\eta \circ U = -\delta\xi - |\xi|_H^2 + \operatorname{trace}(\nabla\eta \circ U) \cdot \nabla\xi.$$

**Corollary 1**   For any cylindrical function $F$ on $W$, we have

$$
\begin{aligned}
E[F \circ V] &= E[F|\Lambda_\xi|]. \\
E[F \circ U] &= E[F|\Lambda_\eta|].
\end{aligned}
$$

**Proof:**   The first part follows from the identity

$$
\begin{aligned}
E[F|\Lambda_\xi|] &= E[F \circ V \circ U |\Lambda_\xi|] \\
&= E[F \circ V].
\end{aligned}
$$

To see the second part, we have

$$E[F \circ U] = \; = \; E\left[F \circ U \frac{1}{|\Lambda_\xi| \circ V} \circ U \, |\Lambda_\xi|\right]$$

$$= \; E\left[F \frac{1}{|\Lambda_\xi| \circ V}\right].$$

From the lemma, it follows that

$$\frac{1}{|\Lambda_\xi| \circ V} \; = \; \frac{1}{|\det_2(I + \nabla\xi) \circ V|} \exp(\delta\xi + 1/2|\xi|^2) \circ V$$

$$= \; \frac{1}{|\det_2(I + \nabla\xi) \circ V|} \exp -\delta\eta - 1/2|\eta|^2 + \text{trace}((\nabla\xi \circ V) \cdot \nabla\eta)$$

$$= \; |\Lambda_\eta|,$$

since, for general Hilbert-Schmidt maps $A$ and $B$, we have

$$\det_2(I + A) \cdot \det_2(I + B) = \exp \text{trace}(AB) \cdot \det_2((I + A)(I + B)),$$

(in fact this identity follows from the multiplicative property of the ordinary determinants and from the formula (e) given in [5], page 1106, Lemma 22) and in our case $(I + \nabla\xi \circ V) \cdot (I + \nabla\eta) = I$. Q.E.D.

**Proof of the theorem:** Let $(h_i; i \in \mathbf{N}) \subset W^*$ be a complete orthonormal basis of $H$. For $n \in \mathbf{N}$, let $V_n$ be the sigma algebra on $W$ generated by $\{\delta h_1, \ldots, \delta h_n\}$, $\pi_n$ be the orthogonal projection of $H$ onto the subspace spanned by $\{h_1, \ldots, h_n\}$. Define

$$\xi_n = E\left[\pi_n P_{1/n} u | V_n\right],$$

where $P_{1/n}$ is the Ornstein-Uhlenbeck semigroup on $W$ with $t = 1/n$. Then $\xi_n \to \xi$ in $D_{p,1}(H)$ for any $p > 1$ (cf., Lemma 1 of Chapter VIII). Moreover $\xi_n$ has the following form:

$$\xi_n = \sum_{i=1}^{n} \alpha_i^n(\delta h_1, \ldots, \delta h_n) h_i,$$

where $\alpha_i^n$ are $C^\infty$-functions as explained in the Proposition 3 of Chapter IV. We have

$$\nabla\xi_n = E\left[\pi_n \otimes \pi_n e^{-1/n} P_{1/n} \nabla u | V_n\right],$$

hence

$$\|\nabla\xi_n\| \leq e^{-1/n} E\left[P_{1/n} \|\nabla u\| | V_n\right],$$

and the same inequality holds also with the Hilbert-Schmidt norm. Consequently, we have

$$\|\nabla\xi_n\| \leq c \,, \; \|\nabla\xi_n\|_2 \leq d \,,$$

$\mu$-almost surely and hence, each $\xi_n$ satisfies the hypothesis of Lemma 1. Let us denote by $\eta_n$ the shift corresponding to the inverse of $U_n = I + \xi_n$ and let $V_n = I + \eta_n$. Denote by $\Lambda_n$ and $L_n$ the densities corresponding, respectively, to $\xi_n$ and $\eta_n$, i.e., with the old notations

$$\Lambda_n = \Lambda_{\xi_n} \text{ and } L_n = \Lambda_{\eta_n}.$$

We will prove that the sequences of densities

$$\{\Lambda_n : n \in \mathbf{N}\} \text{ and } \{L_n : n \in \mathbf{N}\}$$

are uniformly integrable. In fact we will do this only for the first sequence since the proof for the second is very similar to the proof of the first case. To prove the uniform integrability, from the lemma of de la Vallé-Poussin, it suffices to show

$$\sup_n E\left[|\Lambda_n| |\log \Lambda_n|\right] < \infty,$$

which amounts to show, from the Corollary 1, that

$$\sup_n E\left[|\log \Lambda_n \circ V_n|\right] < \infty.$$

Hence we have to control

$$E\left[|\log \det{}_2(I + \nabla \xi_n \circ V_n)| + |\delta \xi_n \circ V_n| + 1/2|\xi_n \circ V_n|^2\right].$$

From the Lemma 2, we have

$$\delta \xi_n \circ V_n = -\delta \eta_n - |\eta_n|_H^2 + \text{trace}(\nabla \xi_n \circ V_n) \cdot \nabla \eta_n,$$

hence

$$
\begin{aligned}
E[|\delta \xi_n \circ V_n|] &\leq \|\delta \eta_n\|_{L^2(\mu)} + E[|\eta_n|^2] + E[\|\nabla \xi_n \circ V_n\|_2 \|\nabla \eta_n\|_2] \\
&\leq \|\eta_n\|_{L^2(\mu,H)} + \|\eta_n\|_{L^2(\mu,H)}^2 + \|\nabla \eta_n\|_{L^2(\mu,H \otimes H)} + \frac{d^2}{1-c} \\
&\leq \|\eta_n\|_{L^2(\mu,H)} + \|\eta_n\|_{L^2(\mu,H)}^2 + \frac{d(1+d)}{1-c},
\end{aligned}
$$

where the second inequality follows from

$$\|\delta \gamma\|_{L^2(\mu)} \leq \|\nabla \gamma\|_{L^2(\mu,H \otimes H)} + \|\gamma\|_{L^2(\mu,H)}.$$

From the Corollary 1 of Chapter VIII, we have

$$\sup_n E[\exp \alpha |\eta_n|^2] < \infty,$$

for any $\alpha < \frac{(1-c)^2}{2d^2}$, hence

$$\sup_n E[|\eta_n|^2] < \infty.$$

We have a well-known inequality (cf.,[5], page 1106), which says that

$$|\det{}_2(I + A)| \leq \exp\frac{1}{2}\|A\|_2^2$$

for any Hilbert-Schmidt operator $A$ on $H$, applying this to our case, we obtain

$$\sup_n |\log\det{}_2(I + \nabla\xi_n \circ V_n)| \leq \frac{d^2}{2}$$

and this proves the uniform integrability.

Since the sequence $(\Lambda_n; n \in \mathbf{N})$ is uniformly integrable, it converges to $\Lambda_u$ in $L^1(\mu)$, hence we have

$$E[F \circ T \,|\Lambda_u|] = E[F],$$

for any $F \in C_b(W)$, where $T(w) = w + u(w)$.

To show the convergence of the inverse flow we have:

$$
\begin{aligned}
|\eta_n - \eta_m| &\leq |\xi_n \circ V_n - \xi_m \circ V_n| + |\xi_m \circ V_n - \xi_m \circ V_m| \\
&\leq |\xi_n \circ V_n - \xi_m \circ V_n| + c|\eta_n - \eta_m|,
\end{aligned}
$$

since $c < 1$, we obtain:

$$(1 - c)|\eta_n - \eta_m| \leq |\xi_n \circ V_n - \xi_m \circ V_n|.$$

Consequently, for any $K > 0$,

$$
\begin{aligned}
\mu\{|\eta_n - \eta_m| > K\} &\leq \mu\{|\xi_n \circ V_n - \xi_m \circ V_n| > (1 - c)K\} \\
&= E\left[|\Lambda_n| 1_{\{|\xi_n - \xi_m| > (1-c)K\}}\right] \to 0,
\end{aligned}
$$

as $n$ and $m$ go to infinity, by the uniform integrability of $(\Lambda_n; n \in \mathbf{N})$ and by the convergence in probability of $(\xi_n; n \in \mathbf{N})$. As the sequence $(\eta_n; n \in \mathbf{N})$ is bounded in all $L^p$ spaces, this result implies the existence of an $H$-valued random variable, say $v$ which is the limit of $(\eta_n; n \in \mathbf{N})$ in probability. By uniform integrability, the convergence takes place in $L^p(\mu, H)$ for any $p > 1$ and since the sequence $(\nabla\eta_n; n \in \mathbf{N})$ is bounded in $L^\infty(\mu, H \otimes H)$, also the convergence takes place in $D_{p,1}(H)$ for any $p > 1$. Consequently, we have

$$E[F(w + v(w)) \,|\Lambda_v|] = E[F],$$

and

$$E[F(w + v(w))] = E[F \,|\Lambda_u|],$$

for any $F \in C_b(W)$.

Let us show that $S : W \to W$, defined by $S(w) = w + v(w)$ is the inverse of $T$ : let $a > 0$ be any number, then

$$
\begin{aligned}
\mu\{\|T \circ S(w) - w\|_W > a\} &= \mu\{\|T \circ S - U_n \circ S\|_W > a/2\} \\
&\quad + \mu\{\|U_n \circ S - U_n \circ V_n\|_W > a/2\} \\
&= E[|\Lambda_u| 1_{\{\|T - U_n\|_W > a/2\}}] \\
&\quad + \mu\{|\xi_n(w + v(w)) - \xi_n(w + \eta_n(w))| > a/2\} \\
&\leq E[|\Lambda_u| 1_{\{|u - \xi_n| > a/2\}}] \\
&\quad + \mu\{|v - \eta_n| > \frac{a}{2c}\} \to 0,
\end{aligned}
$$

as $n$ tends to infinity, hence $\mu$-almost surely $T \circ S(w) = w$. Moreover

$$
\begin{aligned}
\mu\{\|S \circ T(w) - w\|_W > a\} &= \mu\{\|S \circ T - S \circ U_n\|_W > a/2\} \\
&\quad + \mu\{\|S \circ U_n - V_n \circ U_n\|_W > a/2\} \\
&\leq \mu\{|u - \xi_n| > \frac{a(1-c)}{2c}\} \\
&\quad + E[|\Lambda_{\eta_n}| 1_{\{|v - \eta_n| > a/2\}}] \to 0,
\end{aligned}
$$

by the uniform integrability of $(\Lambda_{\eta_n}; n \in \mathbf{N})$, therefore $\mu$-almost surely, we have $S \circ T(w) = w$.                                                                               Q.E.D.

# Bibliography

[1] H. Airault and P. Malliavin: "Intégration géométrique sur l'espace de Wiener". Bull. Sci. Math. Vol. 112 (1988).

[2] J.-M. Bismut: "Martingales, the Malliavin calculus and hypoellipticity under general Hörmander conditions". Zeit. Wahr. verw. Geb. 56, p.469-505 (1981).

[3] N. Bouleau and F. Hirsch: Dirichlet Forms and Analysis on Wiener Space. De Gruyter Studies in Math., Vol. 14, Berlin-New York, 1991.

[4] L. Decreusefond, Y. Hu and A. S. Üstünel: "Une inégalité d'interpolation sur l'espace de Wiener". CRAS, Paris, Série I, Vol. 317, p.1065-1067 (1993).

[5] N. Dunford and J. T. Schwarz: Linear Operators, vol. II. Interscience, 1957.

[6] D. Feyel: "Transformations de Hilbert-Riesz". CRAS. Paris, t.310, Série I, p.653-655 (1990).

[7] B. Gaveau and P. Trauber: "l'Intégrale stochastique comme opérateur de divergence dans l'espace fonctionnel". J. Funct. Anal. 46, p.230-238 (1982).

[8] K. Ito and M. Nisio: "On the convergence of sums of independent Banach space valued random variables". J. Math., 5, Osaka, p. 35-48 (1968)

[9] H. Körezlioglu and A. S. Üstünel: "A new class of distributions on Wiener spaces". Procedings of Silivri Conf. on Stochastic Analysis and Related Topics, p.106-121. Lecture Notes in Math. Vol. 1444. Springer, 1988.

[10] P. Krée: "Continuité de la divergence dans les espaces de Sobolev relatifs à l'espace de Wiener". CRAS, 296, p. 833-834 (1983).

[11] S. Kusuoka, "The nonlinear transformation of Gaussian measures on Banach space and its absolute continuity, I", J. Fac. Sci. Univ. Tokyo, Sect. IA, Math. 29, pp. 567–598 (1982).

[12] P. Malliavin: "Stochastic calculus of variations and hypoelliptic operators". In International Symp. SDE Kyoto, p.195-253, Kinokuniya, Tokyo, 1978.

[13] P. A. Meyer: "Notes sur les processus d'Ornstein-Uhlenbeck". Séminaire de Probabilités XVI, p. 95-133. Lecture Motes in Math. Vol. 920. Springer, 1982.

[14] J. Neveu: "Sur l'espérance conditionnelle par rapport à un mouvement brownien". Ann. Inst. Henri Poincaré, (B), Vol. XII, p. 105-110 (1976).

[15] D. Nualart: "Markov fields and transformations of the Wiener measure". In Stochastic Analysis and Related Topics, Proceedings of Fourth Oslo-Silivri Workshop on Stochastic Analysis, T. Lindstrøm, B. Øksendal and A. S. Üstünel (Eds.), p. 45-88 (1993). Gordon and Breach, Stochastic Monographs, Vol. 8.

[16] D. Nualart and A.S. Üstünel: "Mesures cylindriques et distributions sur l'espace de Wiener". In Proceedings of Trento meeting on SPDE, G. Da Prato and L. Tubaro (Eds.). Lecture Notes in Math., Vol. 1390, p.186-191. Springer, 1989.

[17] D. Ocone: "Malliavin calculus and stochastic integral representation of functionals of diffusion processes". Stochastics 12, p.161-185 (1984).

[18] G. Pisier: "Probabilistic Methods in the Geometry of Banach Spaces". In Probability and Analysis, p.167-241. Lecture Notes in Math. Vol. 1206. Springer, 1986.

[19] R. Ramer, "On linear transformations of Gaussian measures", J. Funct. Anal., Vol. 15, pp. 166-187 (1974).

[20] D. W. Stroock: "Homogeneous chaos revisited". Séminaire de Probabilités XXI, p. 1-8. Lecture Notes in Math. Vol. 1247. Springer, 1987..

[21] A. S. Üstünel: "Representation of distributions on Wiener space and Stochastic Calculus of Variations". Jour. Funct. Analysis, Vol. 70, p. 126-139 (1987).

[22] A. S. Üstünel: "Intégrabilité exponentielle de fonctionnelles de Wiener". CRAS, Paris, Série I, Vol. 315, p.279-282 (1992).

[23] A. S. Üstünel: "Exponential tightness of Wiener functionals". In Stochastic Analysis and Related Topics, Proceedings of Fourth Oslo-Silivri Workshop on Stochastic Analysis, T. Lindstrøm, B. Øksendal and A. S. Üstünel (Eds.), p. 265-274 (1993). Gordon and Breach, Stochastic Monographs, Vol. 8.

[24] A. S. Üstünel: "Some exponential moment inequalities for the Wiener functionals". Preprint, 1993.

[25] A. S. Üstünel: "Some comments on the filtering of diffusions and the Malliavin Calculus". Procedings of Silivri Conf. on Stochastic Analysis and Related Topics, p.247-266. Lecture Notes in Math. Vol. 1316. Springer, 1988.

[26] A. S. Üstünel: "Construction du calcul stochastique sur un espace de Wiener abstrait". CRAS, Série I, Vol. 305, p. 279-282 (1987).

[27] A. S. Üstünel and M. Zakai: "On independence and conditioning on Wiener space". Ann. of Proba. Vol.17, p.1441-1453 (1989).

[28] A. S. Üstünel and M. Zakai: "On the structure of independence". Jour. Func. Analysis, Vol.90, p.113-137 (1990).

[29] A. S. Üstünel and M. Zakai: "Transformations of Wiener measure under anticipative flows". Proba. Theory Relat. Fields 93, p.91-136 (1992).

[30] A. S. Üstünel and M. Zakai: "Applications of the degree theorem to absolute continuity on Wiener space". Probab. Theory Relat. Fields, vol. 95, p. 509-520 (1993).

[31] A. S. Üstünel and M. Zakai: "Transformation of the Wiener measure under non-invertible shifts". Probab. Theory Relat. Fields, vol. 99, p. 485-500 (1994).

[32] S. Watanabe: Stochastic Differential Equations and Malliavin Calculus. Tata Institute of Fundemental Research, Vol. 73. Springer, 1984.

[33] A. Zygmund: Trigonometric Series. Second Edition, Cambridge University Press, 1959.

# Subject Index

# Notations

# Springer-Verlag
# and the Environment

We at Springer-Verlag firmly believe that an international science publisher has a special obligation to the environment, and our corporate policies consistently reflect this conviction.

We also expect our business partners – paper mills, printers, packaging manufacturers, etc. – to commit themselves to using environmentally friendly materials and production processes.

The paper in this book is made from low- or no-chlorine pulp and is acid free, in conformance with international standards for paper permanency.

# Lecture Notes in Mathematics

For information about Vols. 1–1431
please contact your bookseller or Springer-Verlag

Vol. 1566: B. Edixhoven, J.-H. Evertse (Eds.), Diophantine Approximation and Abelian Varieties. XIII, 127 pages. 1993.

Vol. 1567: R. L. Dobrushin, S. Kusuoka, Statistical Mechanics and Fractals. VII, 98 pages. 1993.

Vol. 1568: F. Weisz, Martingale Hardy Spaces and their Application in Fourier Analysis. VIII, 217 pages. 1994.

Vol. 1569: V. Totik, Weighted Approximation with Varying Weight. VI, 117 pages. 1994.

Vol. 1570: R. deLaubenfels, Existence Families, Functional Calculi and Evolution Equations. XV, 234 pages. 1994.

Vol. 1571: S. Yu. Pilyugin, The Space of Dynamical Systems with the $C^0$-Topology. X, 188 pages. 1994.

Vol. 1572: L. Göttsche, Hilbert Schemes of Zero-Dimensional Subschemes of Smooth Varieties. IX, 196 pages. 1994.

Vol. 1573: V. P. Havin, N. K. Nikolski (Eds.), Linear and Complex Analysis – Problem Book 3 – Part I. XXII, 489 pages. 1994.

Vol. 1574: V. P. Havin, N. K. Nikolski (Eds.), Linear and Complex Analysis – Problem Book 3 – Part II. XXII, 507 pages. 1994.

Vol. 1575: M. Mitrea, Clifford Wavelets, Singular Integrals, and Hardy Spaces. XI, 116 pages. 1994.

Vol. 1576: K. Kitahara, Spaces of Approximating Functions with Haar-Like Conditions. X, 110 pages. 1994.

Vol. 1577: N. Obata, White Noise Calculus and Fock Space. X, 183 pages. 1994.

Vol. 1578: J. Bernstein, V. Lunts, Equivariant Sheaves and Functors. V, 139 pages. 1994.

Vol. 1579: N. Kazamaki, Continuous Exponential Martingales and *BMO*. VII, 91 pages. 1994.

Vol. 1580: M. Milman, Extrapolation and Optimal Decompositions with Applications to Analysis. XI, 161 pages. 1994.

Vol. 1581: D. Bakry, R. D. Gill, S. A. Molchanov, Lectures on Probability Theory. Editor: P. Bernard. VIII, 420 pages. 1994.

Vol. 1582: W. Balser, From Divergent Power Series to Analytic Functions. X, 108 pages. 1994.

Vol. 1583: J. Azéma, P. A. Meyer, M. Yor (Eds.), Séminaire de Probabilités XXVIII. VI, 334 pages. 1994.

Vol. 1584: M. Brokate, N. Kenmochi, I. Müller, J. F. Rodriguez, C. Verdi, Phase Transitions and Hysteresis. Montecatini Terme, 1993. Editor: A. Visintin. VII. 291 pages. 1994.

Vol. 1585: G. Frey (Ed.), On Artin's Conjecture for Odd 2-dimensional Representations. VIII, 148 pages. 1994.

Vol. 1586: R. Nillsen, Difference Spaces and Invariant Linear Forms. XII, 186 pages. 1994.

Vol. 1587: N. Xi, Representations of Affine Hecke Algebras. VIII, 137 pages. 1994.

Vol. 1588: C. Scheiderer, Real and Étale Cohomology. XXIV, 273 pages. 1994.

Vol. 1589: J. Bellissard, M. Degli Esposti, G. Forni, S. Graffi, S. Isola, J. N. Mather, Transition to Chaos in Classical and Quantum Mechanics. Montecatini Terme, 1991. Editor: S. Graffi. VII, 192 pages. 1994.

Vol. 1590: P. M. Soardi, Potential Theory on Infinite Networks. VIII, 187 pages. 1994.

Vol. 1591: M. Abate, G. Patrizio, Finsler Metrics – A Global Approach. IX, 180 pages. 1994.

Vol. 1592: K. W. Breitung, Asymptotic Approximations for Probability Integrals. IX, 146 pages. 1994.

Vol. 1593: J. Jorgenson & S. Lang, D. Goldfeld, Explicit Formulas for Regularized Products and Series. VIII, 154 pages. 1994.

Vol. 1594: M. Green, J. Murre, C. Voisin, Algebraic Cycles and Hodge Theory. Torino, 1993. Editors: A. Albano, F. Bardelli. VII, 275 pages. 1994.

Vol. 1595: R.D.M. Accola, Topics in the Theory of Riemann Surfaces. IX, 105 pages. 1994.

Vol. 1596: L. Heindorf, L. B. Shapiro, Nearly Projective Boolean Algebras. X, 202 pages. 1994.

Vol. 1597: B. Herzog, Kodaira-Spencer Maps in Local Algebra. XVII, 176 pages. 1994.

Vol. 1598: J. Berndt, F. Tricerri, L. Vanhecke, Generalized Heisenberg Groups and Damek-Ricci Harmonic Spaces. VIII, 125 pages. 1995.

Vol. 1599: K. Johannson, Topology and Combinatorics of 3-Manifolds. XVIII, 446 pages. 1995.

Vol. 1600: W. Narkiewicz, Polynomial Mappings. VII, 130 pages. 1995.

Vol. 1601: A. Pott, Finite Geometry and Character Theory. VII, 181 pages. 1995.

Vol. 1602: J. Winkelmann, The Classification of Three-dimensional Homogeneous Complex Manifolds. XI, 230 pages. 1995.

Vol. 1603: V. Ene, Real Functions – Current Topics. XIII, 310 pages. 1995.

Vol. 1604: A. Huber, Mixed Motives and their Realization in Derived Categories. XV, 207 pages. 1995.

Vol. 1605: L. B. Wahlbin, Superconvergence in Galerkin Finite Element Methods. XI, 166 pages. 1995.

Vol. 1606: P.-D. Liu, M. Qian, Smooth Ergodic Theory of Random Dynamical Systems. XI, 221 pages. 1995.

Vol. 1607: G. Schwarz, Hodge Decomposition – A Method for Solving Boundary Value Problems. VII, 155 pages. 1995.

Vol. 1608: P. Biane, R. Durrett, Lectures on Probability Theory. VII, 210 pages. 1995.

Vol. 1609: L. Arnold, C. Jones, K. Mischaikow, G. Raugel, Dynamical Systems. Montecatini Terme, 1994. Editor: R. Johnson. VIII, 329 pages. 1995.

Vol. 1610: A. S. Üstünel, An Introduction to Analysis on Wiener Space. X, 95 pages. 1995.